EXPLAINING SCIENCE'S SUCCESS

ACUMEN RESEARCH EDITIONS

Dale Jacquette: *Logic and How it Gets That Way*
Maria Kronfeldner: *Darwinian Creativity and Memetics*
John Wright: *Explaining Science's Success*

Explaining science's success
Understanding how scientific knowledge works

John Wright

Routledge
Taylor & Francis Group

LONDON AND NEW YORK

First published 2013 by Acumen

Published 2014 by Routledge
2 Park Square, Milton Park, Abingdon, Oxon OX14 4RN
711 Third Avenue, New York, NY 10017, USA

First issued in paperback 2017

Routledge is an imprint of the Taylor & Francis Group, an informa business

Notices
Practitioners and researchers must always rely on their own experience
and knowledge in evaluating and using any information, methods,
compounds, or experiments described herein. In using such information
or methods they should be mindful of their own safety and the safety of
others, including parties for whom they have a professional responsibility.

To the fullest extent of the law, neither the Publisher nor the authors,
contributors, or editors, assume any liability for any injury and/or damage
to persons or property as a matter of products liability, negligence or
otherwise, or from any use or operation of any methods, products,
instructions, or ideas contained in the material herein.

ISBN 13: 978-1-138-10820-2 (pbk)
ISBN 13: 978-1-84465-532-8 (hbk)

British Library Cataloguing-in-Publication Data
A catalogue record for this book is available from the British Library.

Typeset in Minion Pro by JS Typesetting Ltd, Porthcawl, Mid Glamorgan.

Contents

CHAPTER 1
Some surprising phenomena

Paul Feyerabend famously asked, "What's so great about science?" (1976: 310). One possible answer is that it has been surprisingly successful in getting things right about the natural world. And, very plausibly, a reason why science has been of interest to philosophers is because it seems to have been *more* successful in doing this than non-scientific or pre-scientific systems, or religion, or philosophy itself. We seem, moreover, to be able to point to some *general ways* in which science has been surprisingly successful in getting things right. Here are three such:

1. Scientists have formulated some theories that have successfully predicted *novel observations*.
2. Scientists have produced theories about parts of reality that were not observable or accessible at the time those theories were first advanced, but the claims about those inaccessible areas have since turned out to be true.
3. Scientists have, on occasion, advanced on more or less *a priori* grounds theories that subsequently turned out to be highly empirically successful.

Of course, it may be disputed whether these phenomena really are genuine. But if they are genuine, they are surprising, and therefore require explanation. The main aim of this book is to offer an explanation of these phenomena. The aim of the present chapter is to argue that there are phenomena here that genuinely do require explanation.

THE FIRST PHENOMENON

The first of the phenomena is that scientists have formulated theories that have subsequently turned out to have novel predictive success. Observational predictions made by scientific theories are, roughly speaking, of two kinds: predictions of observations that are of the *same kind* as those on the basis of which the theory was initially formulated and predictions that are of a new or different kind from those initially used. Suppose a scientist is studying magnets. The scientist may note that all magnets observed so far have two "poles", a north and a south pole. This leads to the conjecture that all magnets have both a north pole and a south pole. And this conjecture in turn leads to the prediction that the next magnet to be observed will have both a north and a south pole. But this is not a novel prediction. The observation that the next magnet has a north and a south pole is just an observation of the same kind as that which has already been made with previous magnets. We will call it an example of *familiar* predictive success.

There are some predictions, however, that seem to be *novel*. One was the prediction that bringing enough uranium-235 into very close proximity will produce an explosion "brighter than a thousand suns".[1] This certainly seems like a novel prediction: no sequence of events of this sort had ever been observed before; it was, moreover, a prediction that turned out to be successful. It is an example of what we would be inclined to call *novel* predictive success. Another example concerns the behaviour of light. The eighteenth-century scientist Poisson showed that it was a consequence of Fresnel's wave theory of light that if a perfectly round object is placed in a beam of light, the resulting round shadow will have a small white spot in its centre. This phenomenon had, as far as was known at the time, never been observed before, and so it had not been used as evidence for a theory of light. But when the experiment was performed, the shadow was observed as predicted. Yet another example comes from the special theory of relativity. This theory predicted that if two extremely accurate clocks were first synchronized, one of them moved a sufficient distance while the other remained stationary, and then the two reunited, the one that had remained stationary would be slightly ahead of the other. Again, this sort of phenomenon had not been earlier observed, and so had not been used to first construct the special theory. But when an experiment of this sort was performed, the clock that had remained stationary did turn out to be ahead of the other, in agreement with the amount predicted by the theory (Hafele & Keating 1972).[2]

Some more examples[3] of novel predictive success are:

- The prediction of the observable chemical behaviour of the "transuranium" elements. These are artificial elements, produced in particle accelerators. They contain nuclei that are larger than the nucleus of the uranium atom, which is the largest atomic nucleus known to occur naturally. Since these elements are artificial, their behaviour had not been observed prior to their creation. Therefore predictions about their chemical behaviour count as *novel* predictions. But scientists *successfully* predicted their chemical behaviour.
- A related case of novel success is the prediction, from gaps in the periodic table, of hitherto unknown elements and their chemical properties.
- The prediction, by Wolfgang Pauli in 1925, of the existence of the neutrino and the subsequent confirmation of its existence in experiments performed by Cowan and Reines in 1951.[4]
- The prediction made by the general theory of relativity, published by Einstein in 1915, that the observed position of a star would be deflected from its actual position by a powerful gravitational field, and the subsequent confirmation of this prediction in 1919 by Eddington during a solar eclipse.[5]

- The prediction by the general theory of relativity that, due to "time dilation" caused by the strong gravitational field of the Sun, light emanating from the Sun would appear to be shifted towards the red end of the spectrum, and subsequent confirmations of this effect.[6]
- The prediction of observational results confirming the existence of the W and Z particles as a consequence of the theory that the weak subatomic force and the electromagnetic force were just different manifestations of a single underlying force, and the subsequent confirmation of the existence of these particles in experiments performed at CERN in 1983.[7]

All of these examples are cases in which a theory successfully predicted phenomena which are, intuitively at least, different from any of those on the basis of which it had initially been formulated. It seems to be beyond serious dispute that science does, *at least sometimes*, succeed in making novel predictions. So, in what follows, it will be assumed that this phenomenon is genuine.

Of course, one serious problem is: can the notion of a novel prediction be precisely defined and is it, for our purposes, necessary to do so? This question is considered in Chapter 4. For the moment, however, we will simply note that, at an intuitive level, it seems that there are indeed cases of novel predictive success.

THE EXPLANATION OF FAMILIAR PREDICTIVE SUCCESS AND THE EXPLANATION OF NOVEL PREDICTIVE SUCCESS

Examples of familiar predictive success are common and easy to obtain. I note that whenever I have started my lawnmower, the neighbour's dog has started to bark; I then predict that the dog will bark next time I start my mower, and this prediction proves to be successful. I note that whenever my car has frost on it, it is hard to start; I predict the next time it has frost on it, it will again be hard to start, and this prediction is successful. And so on. Predictive successes of this sort occur very frequently. How do we *explain* familiar predictive successes? Presumably, the explanation will be along the following lines:

> The world contains certain regularities or uniformities of the form: "Whenever A, then B". If a person has noticed that, in the past, instances of A have been followed by instances of B, and if they consequently come to predict that the next instance of A will be followed by an instance of B, then there is a good chance their prediction will be successful.

Of course, it is only instances of *certain types* of generalizations that lead to predictions that have a good chance of being successful. For example, perhaps every coin I have taken out of my pocket this morning was minted before 1998, but that does not make it likely that the next coin I take out of my pocket will also have been minted before 1998. It seems that only *certain types* of past regularities are likely to persist in to the future. So, if we are to explain *familiar* predictive success, it is important to stipulate that the instances of regularities that have been observed in the past must have been instances of regularities of a *certain type*. The problem of saying just what types of regularities those are is a difficult one: here we will just refer to them as "appropriate" regularities. Provided that the notion of an appropriate type of regularity can be explicated, the explanation of familiar success will be relatively straightforward. But the explanation of novel success is not so easy. Consider, for example, the prediction of the white spot in the middle of a circular shadow. Instances of the regularity "Whenever a round object is placed in a beam of light, there will be a white spot in the middle of the shadow" had never been observed prior to their first derivation from theory. So we cannot use the explanation of *familiar* predictive success to explain why the prediction of the white spot was successful. It seems that it *wasn't* just a new instance of a regularity that had been observed before, but a *new regularity*. An explanation of a different kind is needed if we are to understand how scientists came to make this successful prediction. There is, therefore, a problem with explaining novel predictive success that is not present when we are merely trying to explain familiar predictive success.

On the face of it, cases of novel predictive success are extremely surprising. It is as if someone were to observe that all magnets have north and south poles and to then make the surprisingly unconnected prediction that, for example, the next bird to be observed will have green plumage, and for this prediction to be found to be correct. From a purely empiricist point of view, it certainly seems very odd that a series of observations in one domain should lead to a prediction in another, quite different domain, and for that prediction to subsequently turn out to be right. It is also very puzzling if an instrumentalist interpretation of scientific theories is adopted. It is as surprising as if a tool originally designed to do one job, such as opening tin cans, should also be able to do another, quite different job, such as programming DVD recorders.

There are some philosophers who deny that any special status attaches to novel predictive success. It has, for example, been argued that the novel predictive success of a theory does not support a scientific realist interpretation of that theory, and that neither does it confer any especially high degree of confirmation on the theory. But it is not claimed here that novel predictive support *does* do either of these things. Here it is only claimed that the various forms novel success described *require explanation*, and the inductive explanation that can be given of non-novel success is not, or at least is not obviously, satisfactory for those types of predictive success we are inclined to classify as "novel".

Of course, the phenomenon of novel predictive success is often advanced as a reason for accepting scientific realism, or the doctrine that (mature) scientific theories are true, or approximately true, descriptions of (sometimes unobservable) parts of reality. But if scientific realism is true, it merely raises another puzzle. Our original problem was "How have we managed to hit upon true predictions of types of phenomena not observed at the time of the prediction?" If scientific realism is accepted, we are confronted with the new problem "How have we managed to hit upon *true* descriptions of parts of reality not directly observable?" Whether scientific realism is true or false, it offers a solution to our original problem by postulating another phenomenon which, on the face of it, appears to be at least as puzzling.

THE GENUINENESS OF THE SECOND PHENOMENON

The second phenomenon is the ability of science to produce theories that make true statements about parts of the world that were not accessible to those who formulated those theories.

The claim that this second phenomenon is genuine must be distinguished from the thesis of *scientific realism*, at least as that doctrine is frequently interpreted. A common interpretation of scientific realism is that mature scientific theories are typically (approximately) true. An alternative formulation, which certainly need not be equivalent to the first, is that the terms of mature scientific theories are typically *referential*, or that the entities postulated by typical mature scientific theories *actually exist*. But in either of these formulations, scientific realism is a much stronger thesis than the claim that there are *some* genuine examples of phenomenon 2. Scientific realism is a thesis about *typical* mature scientific theories. It therefore implies, at least, that *most* mature scientific theories are approximately true, or referential, or both. But phenomenon 2 does not make any claim at all about *most* theories; it only says that *some* theories have given us true descriptions of parts of the world inaccessible to those who advanced the theory.

One example of phenomenon 2 concerns the hypothesis of the planet Neptune. It was first suspected that there might be a planet beyond Uranus that was causing perturbations in its orbit. In 1843 John Adams calculated the position of this hypothetical new planet, and three years later Urbain Le Verrier independently did the same. Initially, therefore, Neptune was merely an entity "of theory": not observed, but postulated to explain the behaviour of Uranus. But the existence of Neptune is now beyond serious dispute. Not only is it clearly visible through telescopes, but spacecraft have flown past it, taking photographs of it. A part of reality not accessible to those who first advanced a theory has since become accessible.

Another example of phenomenon 2 is the germ theory of disease. When first advanced, the theory that disease was due to the presence of tiny organisms made reference to inaccessible parts of reality. But now the reality of at least many of those organisms is surely uncontroversial. For example, "amoebic dysentery" is caused by the presence of a certain strain of amoeba in the body. But, although amoebae cannot be seen with the naked eye, the *existence* of amoebae is now surely as little doubted as the existence of the planet Neptune.

5

There are, broadly speaking, two very different grounds for doubt concerning scientific theories. I will call these two grounds for doubt *inductive* grounds and *ontological* grounds. We have inductive grounds for doubting that all crows are black if we have reason to believe that there may exist somewhere some non-black crows. But this doubt need not be accompanied by any scepticism concerning the very existence of crows. Inductive grounds for doubt are always directed towards *generalizations*. More specifically, we have inductive grounds for doubting "All A are B" iff (i.e. if and only if) we have reason to believe there exists an A that is not a B. On the other hand, we have *ontological* grounds for doubt concerning, for example, the theory of quarks if we doubt that there really are any such things as quarks. More generally, some scientific theories make a claim of the following form: "There are some observed phenomena, or there is some accepted fact or state of affairs, that are to be explained by postulating some entity or class of entities *E*." We have ontological grounds for doubt concerning this claim if we have reason to believe that examples of *E* do not exist.[8] Unlike inductive grounds for doubt, ontological grounds need not be exclusively directed towards generalizations. For example, it is possible to have ontological grounds for doubt concerning the hypothesis that perturbations in the orbit of Uranus are due to Neptune, but this need not entail the falsity of any accepted universal generalization.[9]

There is always some degree of inductive doubt about any generalization that is not an analytic truth, although once the right kind of generalization has received just a few positive instances, we usually become pretty confident that *some* universal generalization like it is true. But ontological doubt is different. The controversial region of debate between realists and anti-realists with respect to *ontological* doubt is changing. Here are some things which were once the subject of ontological doubt, but which are no longer (or at least, the extent to which they are now the subject of such doubt is very much less): the (approximate) sphericity of the Earth, the fact that the Earth has a nearly circular orbit, the existence of the rings of Saturn and of the planets Neptune and Pluto, the status of the fixed stars as entities like our Sun, and the existence of galaxies, organic cells, bacteria, viruses, molecules, atoms, protons, electrons and neutrons.

Many theories that were once subject to ontological doubt are now much less so. But this is a phenomenon that requires explanation: how is it that theories which, when initially suggested, postulated entities that were not observable, subsequently turned out to be right about the existence of those entities? This is one type of phenomenon 2.

Of course, the history of science *also* contains many examples of theories that postulated things that turned out not to exist. Some well-known examples are: the crystalline spheres in which the planets were embedded, vortices responsible for the orbits of the planets, tiny screws on the surface of magnets responsible for magnetic force, the "lumeniferous ether", phlogiston and caloric. But the existence of many theories that turned out to be wrong about the less accessible parts of reality does not necessarily remove the need to explain how it is that some other theories got it right. Neither does it remove the need to explain how some theories about less accessible regions went on to receive subsequent confirmation, whether or not the claims they made turned out to be true, or will ultimately turn out to be true.

So, in summary, there do seem to be at least *some* cases of phenomenon 2. But then we are confronted with a problem: how have scientists managed to arrive at true beliefs about parts of nature that were not accessible to them when those beliefs were formulated?

THE GENUINENESS OF THE THIRD PHENOMENON

The third phenomenon is that there are some cases of scientific theories initially advanced on more or less a priori grounds, which nevertheless turn out to be surprisingly empirically successful. Whether there really are cases of this third phenomenon is perhaps rather more controversial than the claim that there are cases of the other two sorts. But there are, I think, some plausible cases. One plausible class of such theories are the conservation laws. The idea that it is something like an *a priori* truth that matter is conserved is reflected in the principle *ex nihilo nihili fit* (from nothing nothing comes). This is claimed as a *necessary* truth in, for example, Aquinas's "Third Way" of establishing the existence of God. Of course, this does not establish that the principle is either genuinely *a priori* or true. But it lends support to the idea that some, at least, have thought of it as something self-evident and not requiring proof. There *seems* at least to be an *a priori plausibility* to the idea that fundamental physical quantities such as mass and energy must be conserved. It has also been claimed that Newton's three laws of motion are *a priori*.[10] In this section it will be argued that there *is* a sense in which these theories do seem to have at least a kind of *a priori* attractiveness or preferability.[11] Nonetheless, they have been surprisingly successful empirically.

First, we need to note that there are (*at least*) three different notions of "*a priori*". On what I take to be the standard notion of "*a priori*", a proposition is knowable *a priori* iff the only experience necessary to know that it is true is that which is necessary to know the meanings of the terms in it. I will call this type of *a prioricity* "*a priori knowledge*" or "*a priori certainty*". But there is a second type of *a prioricity*, which I will call "*a priori preferability*". A proposition is *a priori* preferable iff we find ourselves *inclined* to believe it even though there is no evidence in our experience for its truth. It must be emphasized that no claim is being made that if a belief is *a priori* preferable it is thereby *known* to be true, much less "certain". It is merely to say that we are *disposed* to believe it, despite the fact that there is no empirical evidence for it. Finally, we may also note a third (intermediate) sense of *a prioricity*: a belief is *a priori* in this third sense iff there is no empirical reason to believe it, yet there is still some sort of reason or ground in its support, even if those reasons or grounds are not strong enough to justify calling the *a priori* belief "knowledge".

It will be argued that there are some scientific theories that have enjoyed surprising success that have been *a priori* in at least the second of these senses, that is that we find ourselves *a priori* disposed to accept them.

Let us begin by considering the law of conservation of mass. This states that the total quantity of mass in any closed system is precisely the same before and after any reaction or interaction. In our everyday experience, this law seems to be more or less confirmed. We add the ingredients for a cake together and mix them around, and the mass of the mixed ingredients seems to be more or less the same as the sum of the masses of the individual ingredients. But note that although our everyday experience is *compatible* with the mass being the same before and after, it is also compatible with very many other laws. For example, our experience is compatible with the mass being 0.00001 per cent greater, or less, after the mixing. Even if we were to weigh the ingredients before and after with a pair of kitchen scales, the obtained results would be compatible with the mass being some very small amount greater or smaller after the mixing, due to the imperfect accuracy of the scales. So, while our *everyday* experience is compatible with the law of the conservation of mass, it is also compatible with many, indeed an infinite number, of other hypotheses such as "The mass is greater by 0.00001 per cent" or "The mass is greater by 0.000001 per cent".

It might be objected that if things always became, for example, more *massive* after mixings or dissolvings, or interactions of some other kind, then, in the long run, we would notice after many such interactions that things would eventually become noticeably heavier. But it is easy to think of other possible laws that would prevent any increase or decrease in mass ever becoming detectable. Perhaps things slightly increase their mass in the first, third and fifth interactions, while decreasing on the second, fourth and sixth and so on. Or interactions on Mondays, Wednesdays and Fridays bring about a slight but undetectable increase in mass, while those on Tuesdays, Thursdays and Saturdays bring about a decrease. On Sundays, mass is conserved. Again, there is clearly an infinite number of ways in which mass could be slightly increasing or decreasing all the time, and all these different ways are compatible with our everyday experience.

Despite the fact that there are many laws about how matter behaves that are compatible with our everyday experience, we find the idea that it is *conserved* much more plausible than the alternatives, such as that it alternately increases and decreases by an imperceptible amount. "*Why* would it increase? Where would the extra matter come from? How would it arrange itself to increase and decrease *alternately*?" These are the questions we would ask ourselves. And the absence of an answer to these questions would make the alternative hypotheses much less plausible than the idea that matter is conserved. So I take it as uncontroversial that intuitively, or from a common-sense perspective, we would regard the hypothesis that matter is conserved as being much more likely or reasonable than the others.

But now, our conviction that it is more likely that matter is conserved is clearly not supported by anything in our *experience*. While our experience is *compatible* with its being conserved, it is also compatible with very many other hypotheses.

We regard it as more likely that matter is conserved than that it, say, alternately increases and decreases by a very small amount, but this belief is not derived from or indeed given any support by our experience. It seems, therefore, that it is belief we find *a priori* plausible.

We now come to a surprising phenomenon that requires explanation. The hypothesis that matter is *conserved* is accepted despite the fact that there are many other beliefs that are also compatible with our everyday experience. But when we move from *everyday* experience to, for example, the more precise measurements of the chemical laboratory, we find that the hypothesis that matter is conserved continues to be confirmed, even though many of the alternative hypotheses are falsified. Of course, the hypothesis that matter is conserved is not the only hypothesis consistent with the more precise measurements of the chemical laboratory. The hypothesis that, for example, matter alternately increases and decreases by 0.00000000001 per cent might not be ruled out by even the most precise possible measurements. But the hypothesis that matter is conserved is *among* those that remain unrefuted. And so we are confronted with a puzzling phenomenon: why is it that the hypothesis that we regarded as *a priori* preferable continued to receive confirmation when we started to measure reality more accurately, when many other hypotheses, also consistent with our observations up to that point, were refuted? This is an example of phenomenon 3, and is an example of the type of phenomenon with which we will here be concerned.

One type of answer that has had its adherents is that the law of the conservation of mass does not really make a claim about how reality is at all; it is, rather, merely a *conventional* truth. But this point of view is now difficult to take seriously. According to the special theory of relativity, it is false that matter is always conserved: it is sometimes converted into energy. Moreover, it has been observed that there has been an apparent decrease in mass after atomic explosions, and in nuclear reactors. When interpreting these observations, scientists have not concluded that the missing mass must still exist somewhere undetected; instead, they conclude that the mass has not been conserved at all, but has been converted into energy. That scientists interpret the results in this way is incompatible with the idea that the law of conservation is a conventional truth. But if it is not a conventional truth, we cannot use the idea that it *is* to explain why it continues to be confirmed when we pass from everyday observations to observations made in the chemical laboratory. The falsifiability of the law of conservation of mass assures us that it is a synthetic statement with empirical content. But then its confirmation, once we move from everyday experience to the chemical laboratory, becomes all the more surprising. Why should an empirical, synthetic statement which we find *a priori* plausible continue to receive confirmation when we test it at a more precise level, when other statements, equally confirmed by everyday experience but not *a priori* plausible, are refuted? Does this mean *a priori* plausibility actually makes subsequent empirical success more likely? But why should this be so? We are confronted here with a phenomenon that requires explanation.

The above remarks about the law of the conservation of mass also apply to a number of other scientific laws. Most obviously, they also apply to other conservation laws. But they apply also to some laws which *can be interpreted* as consequences of conservation laws. One example of such a law is the inverse square law of gravitation. If we imagine or conceive of gravity as being due to some kind of substance, or stream of particles, that emanates from the massive body that is the source of the gravitational field, then, provided that the total quantity of the substance is conserved as it moves away from the mass, the *density* of the substance will diminish with the inverse square of the distance from the source. Therefore, if gravity is due to something that is *conserved* as we move away from its source, we would expect it to obey an inverse square law. The same, of course, can be said for any force. So, the idea that forces should obey inverse square laws is *a priori* preferable, in the same way as are conservation laws. But, like conservation laws, the hypothesis that gravity obeys an inverse square law has received confirmation in domains that go beyond everyday experience. For example, we find it to be confirmed when we study the motions of the planets, including the outer planets Neptune and Pluto. Another example is Coulomb's law, which asserts that the magnitude of electric force obeys an inverse square law with increasing distance from a point charge. Coulomb himself was able to verify the correctness of this law to an accuracy of about three per cent. Maxwell was able to show that it was correct to one place in about 40,000. It is now known to be accurate to one place in more than one billion (Friedman 1966: 483). While the data available to Coulomb were compatible with an inverse square, they were also compatible with many other laws. But the inverse square law – the one actually postulated by Coulomb – continued to be confirmed as more accurate measurements became available, while most of the others compatible with the observations available to Coulomb would have been refuted.

There are several additional points to note here. First, like the law of the conservation of mass, the inverse square law for gravitation cannot be a conventional truth, because scientists have considered the possibility that it may be false.[12] Moreover, the argument for the *a priori* preferability of inverse square laws does not contain any essential reference to *either* gravity or to electric force: if it were sound, it would apply to any force. But some forces, such as those operating inside the atom, are not believed to be even approximately inverse-square.[13] Therefore, both the inverse square law of gravitation and Coulomb's law, like the law of conservation, are falsifiable, synthetic statements that make empirical claims about how the world is. It is therefore puzzling, and something that requires explanation, that these *a priori* plausible claims should have received confirmations when we came to closely study nature more closely.

Of course, we now believe that gravity is not due to anything like a stream of particles or a fluid emanating from massive objects. We now believe it is due to the "curved" nature of space-time in the vicinity of massive objects. But this does not make it any less surprising that the inverse square law should have been confirmed when we came to study the planets. In fact, it makes it more surprising, since we are now confronted with the case of an *a priori* preferable theory which makes a genuine factual claim about the world, based on a false assumption concerning the mechanism by which gravity works, nonetheless gaining empirical confirmation.

I conclude that there are at least *some* cases of phenomenon 3.

Although the examples of *a priori* preferable theories that we have discussed all come from pre-twentieth-century science, it would be a mistake to think that the preferability of such theories is exclusively a pre-twentieth-century phenomenon. For example, although we no longer believe that mass is conserved, we do believe mass/energy is conserved. But why do we believe mass/energy is *conserved* rather than that it imperceptibly increases or decreases? It is compatible with even our most precise measurements that it slightly increases or decreases. So we believe it is conserved even though there are other hypotheses equally compatible with all our experience. It is true that in quantum theory we do not believe that mass/energy is conserved in individual interactions. But even in quantum theory, the conservation of mass/energy comes out as something that is true in the *statistical limit*. But note that, while our experience is compatible with it being conserved in the statistical limit, it is also compatible with it being very slightly greater or less in the statistical limit. We believe that in the statistical limit it is *conserved* despite the fact that there are other hypotheses compatible with our experience. It seems that even in twentieth- and twenty-first-century science, we still retain an *a priori* preference for certain types of hypotheses.

Of course, *even if it were true* that an *a priori* preference for certain types of scientific theories was an exclusively pre-twentieth-century phenomenon, that would not remove the need to explain why such theories had been successful in the past. We would still be confronted with a surprising phenomenon, even if it existed only before the twentieth century.

THE RELATIONS BETWEEN THE THREE PHENOMENA

The three phenomena with which we will be concerned are closely related. The first concerns the ability of science to predict novel observational regularities. Let us suppose that theory T has been advanced on the basis of the observation of empirical regularities R_1, R_2, ..., R_m and that T predicts another empirical regularity R_n, different from any of those in R_1, R_2, ..., R_m. If observational regularity R_n is subsequently found to exist, then theory T will have enjoyed *novel* predictive success. But now let us suppose that theory T makes another prediction. This prediction concerns not novel phenomena at the observational level, but the existence of some non-observational state of affairs R^*. Since R^* could not have been observed by the scientists who originally formulated T, it follows that T was not first advanced on the basis of observations of R^*. But if it subsequently turns out that R^* does exist, then T will have led us to a knowledge of a non-observational or theoretical state of affairs different from any of those on the basis of which T was initially formulated; that is, T will have furnished us with an instance of phenomenon 2. Hence, phenomena 1 and 2 are instances of the more general phenomenon of scientific theories entailing true statements about parts of the world that had not been observed by those proposing the theories; phenomenon 1 concerns as yet unobserved regularities at the theoretical level, while phenomenon 2 concerns states of affairs that were inaccessible to those proposing the theory, because they were too small, or too distant, or for some other reason.

Phenomenon 1 occurs when a theory, based on N regularities at the observational level, successfully predicts an observational regularity different from any of those initially used in its formulation. An *a priori* theory is one that is based on the observation of no empirical regularities at all. We can therefore regard phenomenon 3 as the special case of phenomenon 1 when $N = 0$.

In summary, the three phenomena are all instances of the same general type of phenomenon: they are all instances of scientific theories implying true statements about the world that, in various ways, go significantly beyond those that were initially used in formulating the theories. All three phenomena, therefore, are cases in which *science produces surprisingly more knowledge, or more true statements about reality, than were initially put in to it.* Science seems, to a surprising degree, to give more out than we put in. The ability of science to do this is surely at the core of what we feel is "so great" about science. Our aim here is to explain how it does it.

CHAPTER 2

Some unsatisfactory explanations of the phenomena

In this chapter we will consider some possible explanations of the three phenomena described in Chapter 1. It will be argued that none of them are satisfactory.

MIGHT THE PREDICTIVE SUCCESSES OF SCIENCE SIMPLY BE DUE TO GOOD LUCK?

The first of the phenomena is the ability of science to successfully predict novel phenomena. It is worth stressing at the outset that the question with which we are here concerned – how is the ability of science to predict novel phenomena to be *explained*? – is distinct from the question of whether novel predictive success has any special epistemic status. While many authors have claimed that the ability of a theory to successfully predict novel phenomena is an especially strong form of confirmation, there have also been dissenters from this view.[1] But the issue is logically distinct from the question "How is the ability to make successful predictions to be explained?"

One natural response is to suggest that perhaps this form of success is simply due to good luck. Certainly, science has had some impressive predictive successes: some of these were listed in the previous chapter. But it has also had many predictive failures. The history of science is largely the history of one theory being advanced, tested, refuted and then replaced by another. According to Karl Popper, this is the way in which science progresses: the refutation, or predictive failure, of theories is an integral part of scientific progress. Since science contains many examples of predictive *failures* as well as successes, perhaps the novel predictive *successes* are, roughly, no more than would be expected by chance.

However, a little arithmetic turns out to be enough to persuade us this is *extremely* unlikely. Let us consider, as an example, the confirmations of a branch of modern physics known as quantum electrodynamics (QED). This theory has had a number of predictions that have received impressive empirical confirmation. Although it is not perfect, the degree of agreement between theoretical prediction and experimental observation is nevertheless extremely close. Here is a summary of some of the empirical confirmations of QED.[2]

1. *Magnetic moment of electron*
   value predicted by QED: $1159652359 \times 10^{-12}$
   value obtained by experiment: $11589652410 \times 10^{-12}$
2. *Value of Lamb shift for the hydrogen atom*
   value predicted by QED: 1057.864
   value obtained by experiment: 1057.893
3. *Value of muonium hyperfine structure*
   value predicted by QED: 4463.293
   value obtained by experiment: 4463.30235
4. *Value of positronium hyperfine structure*
   value predicted by QED: 203.3812
   value obtained by experiment: 203.3849
5. *Value of positronium spectrum*
   value predicted by QED: 8.62514
   value obtained by experiment: 8.6284

It should be noted that this list of the empirical successes of QED is not exhaustive. There have also been found impressive agreements between theory and observation concerning the radiative decay rates of orthopositronium and the fine structure of the helium atom (Drell 1979).

Now, let us consider the *a priori* probability of these agreements between theoretical prediction and experimental finding occurring by chance. We will begin by considering the first result: that for the magnetic moment of the electron. In this case, the disagreement between observation and theory is approximately one part in one hundred million. According to some plausible considerations,[3] this means the *a priori* probability of such agreement is about 2×10^{-8}. By the same reasoning, the *a priori* probabilities of the other results are 6×10^{-6}, 1.8×10^{-6}, 7.4×10^{-6} and 6.6×10^{-5} respectively. Assuming all these results are independent of each other, the probability of all these results being obtained is just the product of the individual probabilities, that is $1,054 \times 10^{-30}$, or approximately 10^{-27}.

Clearly, this probability is very low. We can get an intuitive idea of just how low it is by considering the question "How long would it take to get results this improbable by a process of random guessing?" The answer is that if a billion theorists around the world were each advancing one hypothesis per second, twenty-four hours a day, seven days a week, then the amount of time it would take before the probability became greater than one-half that they should have obtained, by chance, results that agreed this closely with observation is longer than the age of the universe. The idea that the predictive successes of QED are just due to chance must, therefore, be rejected.

It might be objected against the above argument that it is based on the assumption that the five results are *independent* of each other. But perhaps this is not so. All five measurements are tests of the correctness of QED. Perhaps the success of one of the tests increases the probability that the next test will also be passed. However, with respect to a point of view that is agnostic with respect to the truth or falsity of any theoretical claim, the tests *are* independent. Consider the following conditional claim:

> *If* the value of one of the quantities (say the magnetic moment of   (C)
> the electron) is X, then there is some better than chance probability
> that the value of another of the quantities (say the Lamb shift of the
> hydrogen atom) will be Y.

What is the status of this claim? Plainly, it is not an *a priori* or analytic truth. If it is true, and known to be true prior to the empirical determination of the value of the Lamb shift, it can only be because some *theoretical* claim is known to be true. Hence, from a point of view that is agnostic with respect to the truth or falsity of any theoretical claim, the value obtained for one of the quantities (say the magnetic moment of the electron) is independent of the value obtained for any of the other magnitudes. Therefore, so long as we remain agnostic about the truth or falsity of any theoretical claim, the figure of 10^{-27} is an accurate estimate of the probability of obtaining theoretical predictions that agree with that degree of accuracy with experimental findings.

But, of course, we might not be agnostic about the truth of all theoretical claims. We might, for example, assert that C is known to be true, or at least that it is rational to accept that C is true. In such a situation, the results of the five tests would not be independent, so we could not conclude that the probability of obtaining all five results is the extremely low figure of 10^{-27} – although, of course, the probability would still have to be no higher than the probability of the most likely of the results, which is the still rather low figure of 6.6×10^{-5}. But if we assert that the results of the tests are not independent, and that the obtaining of one of the results makes the others more likely, we are now confronted with another problem: how do we know of this relation of dependence? How, for example, might we arrive at some conditional assertion such as C? Since C cannot have been verified by direct empirical confirmation, there seem to be only two other possibilities: either C follows from some deep theoretical claim or else it is arrived at by random guesswork. Plainly, the probability of arriving, by random guesswork, at a set of conditionals asserting relations of dependence between the five results given above is just the same as the probability of obtaining, by chance, the five results themselves, that is 10^{-27}. But if we assert that we have arrived at C by deriving it from some deeper theoretical claim, then we are back to square one; that is we are back to the problem of explaining how we have arrived at a theoretical claim that turns out to have *a priori* improbable empirical consequences that subsequently turn out to be confirmed.

Another objection may be raised to the argument given here for the extreme improbability of the predictive successes of science. It may be pointed out that these five successes represent only a part of the picture. Although QED has not yet been falsified, the history of science is full of cases of theories that led to predictions that turned out to be wrong. If the *only* predictions ever made by science were these five, then the fact that subsequent experimental tests confirmed them to this degree of accuracy would indeed be astonishing, but the fact that there have been innumerable predictive failures makes these five successes much less surprising, and much less improbable.

While this response is, of course, correct as far as it goes, it still leaves the success of QED extremely improbable. Even if there had been a million theorists throughout the world advancing one hypothesis per second concerning the values of the five results ever since QED was first formulated, the chances that they should have hit upon, by chance, values that agree so closely with the actual results is less than 10^{-12}, or less than one chance in a thousand billion. This would still be true even if every other theory that had been advanced in the history of science had enjoyed absolutely no empirical success whatsoever. And, of course, although QED is a highly successful theory, it is only *one* of the highly successful theories that have been advanced throughout the history of science. So however we account for the predictive success of science, it seems it is not satisfactory to say it is just due to good luck. Some other explanation is needed.

COULD SCIENTIFIC REALISM EXPLAIN THE PREDICTIVE SUCCESS?

Scientific realists, of course, hold that the success of theories is to be explained by saying that they are true or at least, in some sense, close to the truth.[4] It is, of course, a highly controversial issue in the philosophy of science whether the best explanation of the predictive success of science is the (approximate) truth of scientific theories. But if we say that the success *is* due to (approximate) truth, we are confronted with a new problem: how have scientists managed to hit upon *true* theories? Scientific theories are in many ways about *inaccessible* parts of the world. Some theories are about entities, such as atoms, mesons and so on, that are too small to see. Other theories are about entities, such as radio waves, which are at least as big as many entities we can see but which cannot be directly detected. Others make claims about objects, such as quasars and black holes, too far away from us to be directly detected with our senses. Still other theories make claims about wholly familiar things, but the claims are of such a character that common sense would have no idea how to go about verifying or falsifying them; and indeed the claims can almost seem nonsensical: for example, the theory of general relativity asserts that gravity is curved space-time. Finally, *all* theories are explanatory, universal generalizations. As such, they make claims about a potential infinity of states of affairs, about distant points of space, and about the distant past and distant future. More generally, scientific theories make claims about relatively *inaccessible* parts or aspects of the world. So if we have managed to make true claims about these less accessible parts of the world, the question arises: how have we managed to do this? If we *have* managed to do this, we are simply confronted with what we have referred to as phenomenon 2. If we use scientific realism to explain phenomenon 1, we must then explain phenomenon 2.

Could we, perhaps, say that we have managed to hit upon (approximately) true theories by chance? It is easy to show, however, that the probability of doing this is certainly going to be less than the probability of hitting upon *correct empirical predictions* by chance. We have seen that the probability of hitting upon, by chance, the confirmed observational predictions of quantum electrodynamics is astronomically low. Therefore the probability of hitting upon, by chance, a theory that has the same *empirical consequences* as quantum electrodynamics is also astronomically low. But the probability of hitting upon a theory that, in addition to having the same empirical consequences as QED, is *also* approximately true will surely be lower than the probability of hitting upon a theory that just has the same empirical consequences as QED (van Fraassen 1980). If we cannot say that we have hit upon empirically satisfactory theories by chance, neither can we say we have hit upon (nearly) true theories by chance.

It has been argued above that the chances of hitting upon an approximately true theory must be lower than the chances of hitting upon correct observational predictions. But this argument may be questioned. Maybe theories are "sparse". The set of all possible *observations* is extremely large, but maybe, by comparison, the set of all possible *theories* is much smaller. If so, then perhaps the chances of hitting upon an (approximately) true theory are actually much higher than those associated with merely hitting upon the correct observational predictions. However, this objection fails. The set of all possible explanatory theories is *at least* as large as that of all possible observations. For any *possible* set of empirical results E, there will be *some* "theory" or set of theoretical statements capable of explaining E, although some of those "theories" may be very complex, ad hoc or highly implausible. For example, any conceivable set of empirical results can be explained by saying God had a specific intention to produce precisely those results. Any other possible set of results could be explained by attributing a different intention to God. For any possible set of results E, there will be a theory attributing the corresponding *intention* to God. Or any possible set of results could be explained by postulating an elaborate conspiracy, or postulating some event in the measuring apparatus, such as a sudden surge of current within the apparatus, rather than any event in the part of nature under study. Or it could be postulated that it is just a brute fact of nature that under particular experimental conditions, results E will be obtained. Hence, the chances of hitting upon a *theory* that is at least approximately true cannot be any higher than the chances of hitting upon, by random guesswork, true observational predictions. And we have seen that the chances of that are astronomically low.

One natural response to this argument is to point out that it includes, among the explanatory theories worthy of consideration, such scientifically unusual hypotheses as "this result was due to an intention of God". Perhaps, it may be objected, if we are allowed to include such unusual hypotheses among the set of all hypotheses worthy of consideration, then the set of all such hypotheses will be at least as large as the set of all possible empirical results, and so the chances of hitting upon a *true* theory will be at least as low as the chances of hitting upon a *correct* set of empirical results. But if we are to restrict the set of hypotheses worthy of serious consideration to some *subset* of all possible hypotheses – to, say, the set of hypotheses that are scientifically "plausible" or simple or non-ad hoc – then maybe the chances of hitting upon a true or approximately true theory are greatly increased.

There are many difficulties[5] that confront this suggestion, but here is just one: suppose that restricting the range of admissible hypotheses to those that were simple, or plausible, or "elegant", *did* increase the chances of our hypotheses being true. Then we would be confronted with a new problem: why is it that plausibility, simplicity and so on increase the chances of truth? And if they do, how did we manage to discover those *particular* kinds of simplicity, plausibility and so on that are indicators of truth – after all, what is plausible to Martians might not be plausible to us. These questions are considered in the next section.

The difficulty noted here for "full-blown" scientific realism also affects more qualified or restricted variants of realism. One currently popular modification of "full-blown" scientific realism is the idea that it is only *those parts of a theory most directly responsible for its success* that are to be claimed to be true, or close to the truth. For example, there is surely considerable plausibility in the idea that we do not need to say that Fresnel's theory of light was true *tout court* in order to explain its successful prediction of the "Poisson spot". Perhaps it is enough to merely say that Fresnel's theory was right in what it said about the wave character of light, but wrong in other areas. By our lights it was, for example, wrong about the existence of the ether. One challenge for this approach is perhaps to give us a general criterion for identifying those parts of a theory "directly responsible" for its novel success. But even if this challenge can be successfully met, still another problem remains: How have scientists managed to *hit upon parts* of theories that are true or close to the truth? If we say, for example, that the only parts of Fresnel's theory that needed to be (nearly) correct were those parts describing the wave structure of light, the question remains: "How did Fresnel manage to get the wave structure of light correct?" Like bolder versions of scientific realism, the suggestion that we only need to say that *parts* of theories are true still attributes to scientists a surprising ability to find out how things "out there" are. It still attributes to them the ability to discover the truth about some theoretical, not directly accessible parts of the world. And, on the face of it, this ability would appear to be, if anything, even more impressive and mysterious as the ability to successfully make novel predictions.

Similar difficulties confront another of the more restricted forms of realism: "structural realism".[6] Broadly speaking, this is the idea that we cannot know "the nature" of theoretical entities, but we can have much better grounded beliefs about the structural features of the way they interact with other entities. Structural realists, such as John Worrall (1989), argue that our knowledge of the structural features of light, for example, has exhibited much greater stability than our knowledge of the "nature" of light.[7]

In whatever way structural realism is developed in detail, it will always be confronted with an *epistemological* problem – at least if it claims to give us an explanation of *novel* predictive success. Let us say that S is the structural feature of the world used in the explanation of some novel prediction N. If S is to explain N, it seems that N must be logically entailed by S (or at least that N is entailed by S together with other background assumptions B). And if S (or S and B) are to entail N, it seems that N must be "implicitly contained" in S (or in S and B). But now we find that our original problem, "How have scientists managed to hit upon theories that enjoy novel predictive success?", is no closer to being solved. If N is implicitly contained in S, then we are faced with the new problem "How have scientists managed to hit upon S?" Given that N is "implicitly contained" in S, this new problem is evidently no easier than our original one.[8]

Similar difficulties confront the use of other alternatives to "full-blown" realism to explain success. One alternative to the "full-blown" scientific realist explanation has been offered by Arthur Fine. Fine suggests we can explain why some theories are successful not by saying they are true, but by saying the world is *as if* they are true (Fine 1986: 161). Whatever the merits of Fine's position as an alternative to scientific realism, it is clear it is of little help in the present context. Under what circumstances is it "as if" a theory is true? Presumably, it will be as if a theory is true only if the theory is predictively successful. But then we are back to our original problem: How have scientists managed to hit upon theories that subsequently enjoyed *novel* predictive success? The problem of explaining how scientists have hit upon theories that are such that the world is "as if" they are true would seem to be no easier than the problem of explaining how scientists have hit upon theories that enjoy novel success. Fine's suggestion takes us back to square one.

A similar anti-Realist explanation of the success of science has been offered by P. Kyle Stanford (2000: 149–79). Stanford says it is not necessary to say a theory T is true in order to explain its success; it is sufficient to say that T is *predictively similar* to the true theory in that domain. But again, whether or not Stanford has offered us a viable alternative to realism, it is clear it is of little help here. We are now confronted with the problem "How have scientists managed to hit upon a theory that is *predictively similar* to the true theory?" If the predictive similarity of T to the true theory does not include novel predictive success, then Stanford's account does not explain the phenomena with which we are here concerned. But if it does, we are simply back to our original problem. Either way, Stanford's suggestion is of no help here.

The general conclusion we may draw is that whether we try to explain the success of science either with "full-blown" realism or with one its more qualified, restricted variants, we are presented with a new problem: "How have scientists hit upon theories that are true, close to the truth, true in part or right about the structure or predictively satisfactory in the right way?" Since this problem would seem to be at least as difficult as that of explaining the phenomena with which we are concerned, realism and its variants would seem to take us no closer to a solution to our problem.

THE META-INDUCTIVE EXPLANATION OF THE SUCCESS
OF SCIENTIFIC METHODS

In this section I will consider what is perhaps the most promising strategy for explaining how we have managed to produce theories that enjoy novel predictive success. This is the "meta-inductive" explanation of the success of our methods.[9] The meta-inductive explanation is as follows:

> Scientists are, all the time, advancing hypotheses. By chance, some of these hypotheses enjoy novel predictive success. It is noticed that the ones that enjoy novel predictive success all have (or mostly have) some property M. This property might be, for example, simplicity, "plausibility", naturalness", etc. Here we will just call it "property M". This leads scientists, when considering how to construct *new* theories, or when considering which of a number of candidate hypotheses to accept, to prefer those with property M. Consequently, future hypotheses will (tend to) have property M. But since property M has been shown, by past experience, to be associated with novel success, it is only to be expected that new hypotheses that have M will also have novel success.

Now, some sort of account like this may very well be partially correct. But as an *explanation* of subsequent predictive success it is unsatisfactory. In order to see why, we first need to get clear on precisely what must be assumed if this explanation is to work. It will be easier to do this if we explicitly spell the argument out in a number of distinct steps:

1. Scientists have discovered, or noted, that those theories *advanced in the past* that enjoyed novel predictive success tended to have property M.
2. Scientists hit upon the hypothesis that theories postulated in the future that have property M will tend to enjoy novel predictive success.
3. So: scientists, in their practice, have tended to favour theories with property M.
4. But theories with property M do indeed (tend to) have novel predictive success.

Therefore: Theories advanced by scientists (tend to) have novel predictive success.

The point that needs to be noted here is that step 4 – the claim that theories with property M do indeed tend to have novel predictive success – plays a *non-redundant* role in the explanation. But if 4 is true, it is a rather surprising fact. On the proposed explanation, scientists may discover M either by a "meta-induction" on past theories, or perhaps just by a stab in the dark. But why should property M *continue* to be a reliable guide to future predictive success when scientists come to theorize about new domains? Not that long ago, our theories were about things on the surface of the Earth – rocks, trees, balls rolling down slopes and so on. And for many centuries before, our theories would have been about such familiar, everyday objects. Early scientists may have hit upon some property M possessed by theories about these familiar objects. And, on the meta-inductive account, this led them to *prefer* theories that have property M when they came to theorize about, for example, areas of space beyond the surface of the Earth, or about the very small, or very high velocities, or very high energies. On the meta-inductive account, it is *because* property M continues to be a reliable indicator of future empirical success in these new domains that our theories in these domains enjoy novel success. But if this correct, it is rather *surprising*: it could have been that property M – whether it be our sense of simplicity, plausibility, mathematical "elegance" or whatever – that we had developed as we theorized about tables and chairs, trees, rocks and rivers, and so on, proved to be quite unreliable as a guide to successful theories about protons, mesons and curved space-time. And so we are confronted with the question "Why should M have *continued* to lead to success in these new domains?" A complete explanation of novel success – particularly in these unfamiliar domains – would need to include some kind of explanation of *why* our sense of theoretical goodness did continue to work so well.

An objection may be raised against the preceding argument. Consider a simple scientific law such as "All silver conducts electricity". We may have initially arrived at this law by observing samples of silver on the Earth. But we are nevertheless confident that if we were to encounter samples of silver on, say, the Moon or Mars, they would still conduct electricity. Moreover, this confidence seems entirely reasonable. So – it might be objected – is it not also reasonable to assume that if the generalization "All theories with M (tend to) have subsequent predictive success" has received positive instances in familiar domains, that it will continue to do so when we come to theorize about unfamiliar domains?[10]

However, compared to typical scientific laws, the claim "All theories with property M (tend to) have subsequent success" is very "queer". This can be brought out if we consider what general type of property M is likely to be. If we are confronted with a number of hypotheses, all of which can explain some body of observations, we use M to select which of them is the best. Property M is therefore likely to be something like *simplicity* or *plausibility* or *mathematical elegance* or some similar concept. But it is a feature of these concepts that they are all closely linked to human tastes or preferences. There is, in practice, no test for plausibility or mathematical elegance other than what we find to be plausible or elegant. Some philosophers have proposed definitions of simplicity, or algorithms for measuring it, but what is used in *scientific practice* is scientists' own intuitive judgements concerning simplicity. Therefore, if we are to explain actual cases of success, we must explain the success of theories that have been selected by scientists exercising their own preferences and intuitive judgements. Moreover, the only test we possess for the adequacy of a given definition of, for example, simplicity is whether or not the measures of comparative simplicity it delivers are in accord with our intuitive judgements. But whether a theory subsequently receives experimental confirmation or not, or whether a theory turns out to be close to the truth or not, would seem to be something quite independent of human tastes and preferences. These things would seem to depend on how the world "out there" actually is. So the generalization "All theories with M (tend to) enjoy subsequent predictive success" asserts there is a correlation between something that seems to be closely linked to human tastes and preferences and how things actually are in the universe "out there". This claim may be true, but if it is true it is certainly very surprising. And if it were true, we would want an explanation of this fact.

Note that the reasons why we would want an explanation of "Property M is correlated with subsequent empirical confirmation (or with approximate truth)" is quite different from the reasons why we would want an explanation of "All silver conducts electricity", or any other typical law. That a property that seems to be so closely linked with human tastes and preferences should be correlated with something we believe to be so independent of the human realm – whether or not a theory passes new tests to which we subject it – seems contrary to our general view of the universe and the place of human beings in it. Our view of the place of human beings in nature would lead us to expect that this should *not* be so. In this respect it is quite unlike typical explanatory laws such as "All silver conducts electricity". If it is true that M is correlated with subsequent confirmation, or with truth-likeness, this is a fact which cries out for further explanation, in a way that typical laws do not.

There are some other difficulties with the meta-inductive explanation of success. One is that it asks us to accept certain factual claims which are rather hard to actually believe. Consider, for example, our preference for smooth curves. It does not seem to be the case that we have acquired our preference for smooth curves by a laborious process of trial and error, that is by first trying out all sorts of snaking and jagged curves, finding out they were not good at successfully predicting new values of the properties under consideration, and then finally hitting upon the smooth type of curves we actually do prefer. Or again, consider our preference for simple explanations. It does not *appear* to be the case that people acquire this preference by first trying all sorts of complicated and ad hoc explanations, finding that such explanations are not very good at leading to successful predictions, and then developing a preference for simple explanations. On the face of it, the meta-inductive account of how we came to acquire our epistemic preferences does not seem to square with our experience.

Another, fundamental, shortcoming of the meta-inductive account is that it obviously cannot explain how we came to have a preference for hypotheses that enjoy *inductive* support. Plainly, we cannot say we *acquired* a preference for hypotheses with inductive support by noting that in the past such hypotheses had gone on to enjoy subsequent predictive success, and that we therefore came to believe that in the future such hypotheses would also enjoy subsequent predictive success. Obviously such an explanation presupposes we already have a tendency to adopt inductively supported hypotheses, and so does not explain how we acquired a tendency to prefer such hypotheses.

And there is another shortcoming for the meta-inductive explanation. So far, we have considered whether the meta-inductive account might be able to explain how we manage to hit upon theories that subsequently turn out to have predictive success. It has been argued that it does not seem to be able to provide a satisfactory explanation of this. But it is also worth noting that it would be quite unable to explain the third of the phenomena with which we are here concerned: the empirical success of some *a priori* preferable theories. On the meta-inductive account, we would have arrived at our *preference* for conservation theories, for example, by a process of trial and error; that is we would have tried a range of different types of theories, some of which had quantities conserved, others of which did not, found conservation theories to generally have a better record for subsequent predictive success, and so developed a preference for conservation theories over others. This is, on the meta-inductive account, why we now prefer conservation theories. But this account has an obvious flaw. As we have already noted, all our experience is compatible with the truth of conservation laws, but it is also compatible with the truth of many other laws which say that quantities are always imperceptibly increasing or decreasing. The meta-inductive account is therefore confronted with a problem: why has this process of trial and error led to a preference for *conservation theories* rather than to a preference for one of the indefinitely many other types of theories that are also compatible with our experience?

Note that it is of no help to the advocate of the meta-inductive account to appeal to the greater *simplicity* of conservation laws. The problem that arises with conservation also arises with simplicity. Suppose it is suggested that we acquired our preference for simple theories by a process of trial and error. This would involve trying all sorts of different theories, some of which are simple, others of which are not, and finding that the simple ones tended to enjoy better rates of subsequent confirmation. The difficulty is that, just as many non-conservative laws are as compatible with our experience as conservation laws, so many *complex* theories are as compatible with our experience as simple theories. Our earlier discussion of conservation laws provides us with some examples of this. The hypothesis that matter is conserved is a lot simpler than the hypothesis that it alternately increases and decreases by 0.0000000000001 per cent. But the latter hypothesis is as compatible with our experience as the claim that it is conserved. So it is at least unclear, on the meta-inductive account, why we should have come to acquire a preference for *simple* theories, rather than some sort of non-simple theory that was also compatible with all our experience.

There is yet another difficulty that confronts the meta-inductive explanation of success. Let us suppose that we have come to accept that it is *simple* theories that have worked in the past. How, then, are to explain the fact that we have a preference for simplicity when selecting new theories? Presumably, it is because we have also made an inductive inference of the following form:

> Simple theories have worked in the past.
> _____
> So: simple theories will work in future.

But now we are confronted with another problem: why is it that we have made *this* ampliative inference rather than some other? There are, after all, an indefinitely large number of other conclusions compatible with the claim that simple theories have worked in the past. Plainly, we cannot *explain* why we have used this particular induction because it is simpler than the others and we prefer to make simple inductions, since it is our preference for simplicity which is the very thing we are attempting to explain.[11]

The problem of *explaining* our preference for straight inductions, or for simplicity, is of course a different problem from that of *justifying* our preference for them. But the two problems encounter similar logical difficulties. A problem for any justification of induction is, of course, to avoid begging the question. But we have seen that a similar problem besets attempts to *explain* our preference for, for example, simple theories by saying that we have performed a meta-induction on our own past practices. Saying that we have arrived at our preferences by such a meta-induction implicitly attributes to us preferences of the very sort we are trying to explain.

I conclude that the meta-inductive explanation of the predictive success of science is not satisfactory.

"EVOLUTIONARY" EXPLANATIONS OF SUCCESS

It might be suggested that some kind of "evolutionary" explanation could be given for the success of science.[12] An evolutionary explanation might be given either at the level of individual theories or at the level of our methods. At the level of individual theories, an evolutionary explanation would go something like this: theories are formulated by a process of blind variation. Some of those theories subsequently turn out to be successful; others do not. In this way, science eventually comes to be populated only by successful theories. However, it is plain that this account does not satisfactorily explain highly improbable novel predictive successes. If theories are formulated by a process of blind variation, how are we to account for such *a priori* improbable new empirical successes, such as those of QED described earlier?

An "evolutionary" account might also be given of how we arrived at our tendency to prefer theories that are simple, "elegant", conservative and so on. However, it is clear that this would simply be a variant of the meta-inductive account of how we came to our methods, and so subject to the same difficulties as we encountered in the previous section.

MIGHT THE "LOGIC OF INVENTION" EXPLAIN HOW
WE HAVE HIT UPON SUCCESSFUL THEORIES?

In recent years very great advances have been made in the "logic of invention". Programs have been developed which enable computers to discern patterns in bodies of data and represent those patterns with mathematical formulae.[13] Of course, the programs that have been developed to date have not enabled computers to repeat such intellectual feats as the creation of the theory of general relativity or quantum theory, but perhaps it is just a matter of time before programs are developed which can achieve such things. This would, of course, be a hugely impressive achievement, but would it *explain* the phenomena with which we are here concerned? It will be argued here that it would not. There are two things such a development would leave unexplained. Suppose we did succeed in writing a program which enabled a computer to model our methods; this in itself would not explain how we ourselves hit upon those methods in the first place, and, more important, neither would it provide us with an explanation of *why* those methods lead to subsequent predictive success. These points can perhaps be made clearer if we consider an extremely simple possible model of what "our methods" might be. Suppose our methods consist merely in accepting the *simplest* theory that can explain some body of data. We might be able to develop a program that enabled a computer to find the simplest theory as effectively and reliably as any human scientist. Let us also assume that the simple theories that the computer found turned out, when tested, to enjoy novel predictive success. Would this provide us with an explanation of the phenomenon of novel predictive success in science? It would not, since there are two important things it would fail to explain. First, it would not explain how or why *we* came to prefer simple theories in the first place. But more important, it would leave us completely in the dark why those simple theories went on to enjoy subsequent novel predictive success. In order to explain why simple theories turn out, on subsequent testing, to be successful, we would need more than an understanding of how the human mind, or a computer, came to discern a simple pattern in a complex mass of data; we would also need an explanation of why theories that had this property of simplicity seemed to subsequently "match up" with the world out there.

Of course, it is natural to protest that the criteria used in selecting theories may be much more complicated, rich and diverse than just "simplicity". But even if it is granted that it is some complex of properties that is used, the problem remains: why is it that theories that have those complex of properties go on to enjoy subsequent success?

It may be objected against this that not all intelligent machines need to rely on the application of a *program* to construct or identify good theories. Perhaps "neural nets" could be trained to identify good or simple theories without using anything like a program or algorithm to do so.[14] But whether simplicity (or some complex of properties) is identified by using an algorithm or the recognitional capacities of a neural net, the basic problem would remain untouched: why is it that theories with this property of simplicity also seem to be surprisingly good at leading us to predictions which subsequently turn out to be correct?

Of course, none of this implies that developments in the "logic of invention" would not have some role to play in a complete explanation of novel predictive success and the other phenomena with which we are here concerned. It may help us understand *how* we select the theories we do select. But it would not explain why those theories are *successful* in the ways we want explained.

COULD WE EXPLAIN THE PHENOMENA BY HYPOTHESIZING THAT NATURE ITSELF IS SIMPLE?

One natural explanation of the ability of science to lead to novel predictive success is that nature itself is simple. If nature itself is simple, then simple theories would seem to stand a better chance of being true, and hence predictively successful. However, the hypothesis that nature is simple only provides us with half an explanation of the phenomena. We would also need an explanation of why we *prefer* simple theories; or, more precisely, we would also require an explanation of why the type of theories we prefer have a property that reflects this general feature of the universe. Is it merely good luck that the type of theories we prefer also reflect this feature of the universe? After all, it seems possible that the type of theories we prefer might *not* have reflected this general feature of the universe. But if we say that we *discovered* that simple theories are more likely to be true, or successful, by a process of trial and error, then we are back to the meta-inductive or evolutionary explanations, the shortcomings of which have already been investigated.

It is appropriate to consider here in a little more detail a feature that any explanation of success must have if it is to be satisfactory. Suppose we adopt as a hypothesis that the universe is simple. As we noted above, this leaves unanswered the question of why we prefer simple theories. But suppose we were also able to produce an explanation of why we preferred simple theories that was quite *independent* of our reasons for thinking that the universe was simple. Suppose we explain our preference for simple theories by appealing to, for example, the fact that they are easier to understand and work with. Then we would be in possession of both (a) an explanation of why we prefer simple theories and (b) an explanation of why simple theories are more likely to enjoy predictive success. But would this give us an explanation of why our theories have been successful? There is clearly a respect in which this explanation would be unsatisfactory. This explanation leaves it as just a fortunate fluke that the type of theories that we find easier to understand and work with are of the very same type as those that reflect this large-scale structural feature of the universe, namely its simplicity. And so, on this account, the reason why our methods are successful ultimately comes down to sheer good luck.

Of course, it may be *true* that the success of our methods is just due to good luck. But saying that it is just due to good luck is not an intellectually satisfactory position if the probability of the agreement is extremely low. So how likely *is* it that the type of theory that we find easier to understand and work with should also reflect a large-scale structural property of the universe? This is clearly a very complex question, but it seems pretty clear that the *a priori* probability of the universe obeying laws that are simple enough to be graspable by the human mind is extremely low. This can be brought out in the following way: consider a list containing all logically possible sets of laws that a universe might obey. Suppose this list to start with the simplest possible laws and that, as we go down the list, they become more and more complicated. Perhaps the human mind can grasp laws that contain up to a million symbols. Clearly, the number of possible laws on the list that contain more than a million symbols will be very much larger than the number of laws containing a million symbols or less. And even if we restrict the list to *finitely* complex laws, there will still be a much larger number of laws on the list with more than a million symbols than there will be with less, since there are many more finite numbers greater than a million than there are that are less. So there will be many more laws on the list that are too complicated to be grasped by the human mind than there will be laws that can be so grasped. This seems to show that the *a priori* probability of the universe obeying laws that are simple enough to even be graspable by the human mind is extremely low. The *a priori* probability of it obeying laws that are so simple we find them easy or convenient to work with would therefore be even lower.

The above considerations seem to show that it would not be an intellectually satisfactory position to assert that we prefer simple theories because they are convenient to work with *and* that they are successful because the universe itself is simple. This explanation leaves it as just a fortunate fluke that the type of theory we prefer also reflects the way the universe is, and the *a priori* probability of this fortunate fluke is extremely low.

Of course, the argument just given does not merely apply against the suggestion that we prefer simple theories because they are convenient. An alternative explanation of why we prefer simple theories might appeal to the theory of evolution. It is perhaps tempting to suggest that organisms with simple models of their environment are at an *evolutionary advantage* because they can make predictions more quickly than organisms with more complex models. But this suggestion, too, has an obvious problem: why should a property of theories than enables believers to calculate *quickly* also be a property that mirrors large-scale features of the universe? This would seem to be just a fortunate fluke.

More generally, *any* explanation of the success of our methods which says that simple theories are successful because the universe itself is simple, but our *preference* for simple theories is due to some quite independent factor, runs up against the problem of explaining why it is that the type of theory we prefer should also reflect this feature of the universe.

In conclusion: the idea that the success of our methods can be explained by saying that the universe is simple is unsatisfactory because it does not explain why we prefer simple theories. But even if it were supplemented by some account of why we prefer simple theories, it would still be unsatisfactory if it left it merely as a fortunate fluke that the type of theory we preferred reflected some large-scale property of the universe.

COULD SCIENCE ITSELF PROVIDE US WITH A SATISFACTORY EXPLANATION OF ITS OWN SUCCESSES?

Perhaps the most obvious of the sciences to look to for an explanation of the ability of the human brain or mind to produce successful theories are psychology and/or cognitive science. But a little reflection shows that these sciences could not provide a *complete* explanation of the phenomena with which we are here concerned. We cannot "see directly" into, for example, the interior of the atom, or the behaviour of mesons or quarks. Neither can we see into the future and see directly what the outcomes of future tests of our theories are going to be. So if we do manage to hit upon theories about, say, the interior of the atom that are true, or if we do manage to hit upon a theory that passes future tests, it is *not* because we have *directly seen* the interior of the atom, or the results of future tests, and selected the right theories on that basis. All that we have access to are data about more directly observable parts of the world that record facts about the past and present. And we use criteria such as simplicity, plausibility, elegance and perhaps many others to select, from among all the theories that are capable of explaining our data, that theory which we think is the best. If we are lucky, the theory that we decide is the best does turn out to get subsequent confirmation: we might get subsequent confirmation that what it says about the interior of the atom is right, or it might pass novel tests to which we subject it in the future. Now, plainly, a *complete* explanation of such successes would need to answer (at least) the following two questions:

> *Question 1* How did the human brain or mind, when presented with some body of data, manage to formulate some theory, or find some pattern in the data, that was also simple, plausible, elegant, and so on?
>
> *Question 2* Why is it that the theory that was simple, elegant, plausible, etc. successfully passed novel tests to which we later subjected it? Or: why did we get subsequent confirmation that the simple, plausible theory was true or close to the truth?

Very plausibly, sciences such as psychology or cognitive science are the appropriate ones to look to when answering the first question. But it is also clear that those sciences, by themselves, will not give us an answer to the second question. To answer the second question we need to explain why those theories we judged to be good (those with simplicity, plausibility, etc.) turned out subsequently to be empirically successful in new ways, or to be (approximately, etc.) true. It seems any science-using explanation of this fact would need to mention not just the theories themselves and the mental processes that gave rise to them, but also those parts of the world that the theories are about. For example, suppose we looked to science for an explanation of why the *simplest* theory of the interior of the atom subsequently passed some novel tests. In order to explain this it would not be enough to give some account of how the human brain managed to arrive at a simple theory that explained some collection of data: we would also need to explain why the predictions made by that simple theory concerning how the interior of the atom behaved turned out to be true. So sciences such as psychology and cognitive science are not sufficient to answer question 2.

Sciences such as psychology and cognitive science would fail to provide us with a *complete* explanation of the success of our methods for the same reason that the "logic of invention" would fail to give us a complete explanation. The *success* of our methods is an agreement or confluence between the results of our methods and the behaviour of the world "out there". A complete explanation of this agreement or confluence could not be obtained by focusing exclusively on what we do, or on what goes on inside us.

But now, could we not use science to provide us with an answer to question 2 by bringing in more scientific theories, for example theories about the interior of the atom? On the face of it, this seems to get us no further than some other suggestions already considered. Suppose theories of psychology or neuroscience explained why we preferred, or how we came to recognize, let us say, atomic theories with some property P (such as simplicity). Theories of particle physics might explain why theories with that property P are more likely to be true. But there would still be one surprising fact left unexplained in this account: why is it that the property P of theories – that psychology or neuroscience explains our tendency to prefer – is the *very same* property as that which particle physics tells us is more likely to be possessed by successful theories? Is this just again a fortunate fluke?

It might be suggested that an evolutionary explanation could be given of why the type of theory we prefer also tends to be successful. But we have already noted numerous difficulties with accounts of this sort. And there is another very general reason for doubting that empirical science itself could explain the success of its own methods. Let us suppose that there is some very general scientific theory G. This theory has, we will assume, been very well confirmed. Suppose also that G implies that theories in other domains would be more likely to enjoy subsequent empirical success, or even be (approximately) true, if they were *simple*. Would this give us a satisfactory explanation of why simple theories were more likely to enjoy empirical success? There is a pervasive reason for scepticism about any such explanation. Obviously, we would be unable to "directly see" that G is true: as with all scientific theories, we must rely on indirect evidence for its truth. Among those of its features which would make us more inclined to accept that it is true would be its simplicity. So, G itself would be a simple theory, and we accept G, in part, *because* it is simple. If G entails that the other theories are more likely to be successful (or true) if they are simple, there is at least the possibility that this is merely a by-product of the fact that G, like all our theories, has been accepted by us *because* it is simple, rather than because the universe really is simple. It may be that G's recommendation of simple theories is just an artefact of its having been chosen by the criteria we use to select all our theories, rather than because it has got the universe right on that point. That is, G might not give us any reason to prefer simple theories that is *independent of the fact that we do, in fact, prefer simple theories.* In such a situation G would not give a satisfactory explanation of *why* simple theories are likely to do better.

There is one more reason for scepticism about the idea that science itself might be able to provide us with an explanation of the success of its own methods. It seems at least unlikely that science could do justice to the apparent *ubiquity* of what we regard as the rational methods of procedure. Consider the following examples of different people looking for the correct explanation of certain phenomena:

- A detective considering all the clues found at the scene of a crime, and trying to find the correct reconstruction of the events at the scene, with a view to determining the identity of the perpetrator of the crime.
- A medical doctor considering the symptoms exhibited by a patient, with a view to determining the disease from which the patient is suffering.
- A geologist examining an area of land, and trying to arrive at an explanation of how its topographical features arose, by natural processes, from an earlier state.

Plausibly, in all of these cases, the detective, the doctor and the geologist will use criteria such as simplicity to arrive at their hypotheses. And this seems to us to be a perfectly sensible or rational thing to do. Moreover, it seems reasonably plausible to say that if they exhibit a preference for simple hypotheses, they are more likely to hit upon explanations that are predictively successful, or even close to the truth, than they would if they preferred highly complex or ad hoc hypotheses. It seems reasonably plausible that a preference for simplicity would be a good guide to truth, or to subsequent success, not just in physics, but in other sciences and areas of inquiry as well. But now, if this is so, it seems unlikely that any scientific theory, no matter how general, could explain this success. For example, it does *not* seem very likely that either quantum theory or the theory of general relativity could furnish us with an explanation of why simple explanations of the clues found at the scene of a crime are more likely to be true than more complex explanations. Scientific theories, no matter how general, would not seem to be general enough to explain the success of our methods in all areas of intellectual inquiry.

The above argument assumes that simplicity, lack of ad hocness and so on are guides to subsequent success in very many different sciences, and this is a claim that would need to be established by empirical investigations. But, despite the fact that the truth of this claim has not been established here, it is, I think, a claim with considerable plausibility. So the above argument is at least a *plausibility argument* against the idea that science itself would not provide us with an explanation of the success of its own methods.

COULD A SOCIOLOGICAL EXPLANATION BE GIVEN
OF THE PHENOMENA?

Many workers in science studies nowadays seek sociological explanations of the phenomena of science. Might the phenomena with which we are here concerned be given a sociological explanation? Barry Barnes (1974) has proposed such an explanation. Barnes has pointed out that there is considerable prestige within the scientific community to having theories that are empirically successful. Hence, he argues, there is considerable social pressure on scientists to produce empirically successful theories, and this is why they do in fact produce such theories. However, Barnes's proposed explanation would not appear to be entirely satisfactory. Here it is important to again draw the distinction between *novel* predictive success and more familiar types of success. A theory has novel predictive success if it successfully predicts phenomena – such as Poisson's white spot – different from those on the basis of which the theory was initially formulated. Scientists can certainly *try* to produce theories that enjoy novel predictive success, but whether or not they actually succeed is not merely to be explained by appealing to the fact that they tried. The fact that the white spot was in fact *found* in the circular shadow is not to be explained merely by saying scientists were trying to produce a successful theory;[15] something more is required if we are to explain why the white spot was *actually found*. Social pressure and prestige can perhaps explain why scientists try to find theories with novel predictive success, but it does not provide a complete explanation of their success.

COULD A "KANTIAN-STYLE" EXPLANATION BE GIVEN
OF SOME OF THE PHENOMENA?

So far we have been primarily concerned with explanations of phenomena 1 and 2. We have not focused on explanations of phenomenon 3. This occurs when an *a priori* plausible theory subsequently proves to have empirical success. This phenomenon is certainly surprising: how can it be that a theory we favour *a priori* should appear to give us factual knowledge of the empirical world? This problem is, of course, like Kant's famous question "How is synthetic *a priori* knowledge possible?" So it might be wondered whether a Kantian-style explanation could be given for at least phenomenon 3. Kant, as we all know, thought the human mind possessed *a priori* certain categories and intuitions, and as a consequence always interpreted experience so it never provided us with counter-instances to fundamental laws such as the conservation of matter and that everything has a cause. Nowadays, of course, we think there must be at least something wrong with Kant's account, because we believe some cases of allegedly synthetic *a priori* knowledge (such as "matter is always conserved" and "every event has a cause") are in fact false. But would a Kantian-style explanation satisfactorily account for even the *empirical successes* of *a priori* plausible theories? It will be argued it would not. Let us return to an example we have already mentioned: the discovery of the planet Neptune. The existence of this planet was first suspected because the orbit of the planet Uranus deviated slightly from that which was predicted by the laws of Newton and the influences of the known planets. Using Newton's laws, Le Verrier and Adams calculated where the new planet should be. When J. G. Galle looked in the spot predicted by Le Verrier, he found the predicted planet. It is hard to see how this success could be explained by saying that the human mind has a tendency to interpret experience in a particular way. What would such an explanation be like? That Galle's *a priori* categories and intuitions caused him to perceive a white spot at the location predicted by Le Verrier when he looked through his telescope? This is not at all plausible. Of course, it may very well be that even if the white spot had not been observed, scientists would still not have given up Newton's laws. But that does not explain why the white spot *was* observed. It fails to explain *the way* in which Newton's theories were confirmed. So, it seems that a Kantian-style explanation of phenomenon 3 is not satisfactory.

THE "EXPERIMENTALIST'S REGRESS" AND THE EXPLANATION OF THE PREDICTIVE SUCCESS OF SCIENCE

Recently, the "experimentalist's regress" has formed the focus of much discussion in the history and philosophy of science (Collins 1985). The experimentalist's regress can lead to the idea that the predictive success of science might be nothing more than an artefact of the way we check that our apparatus is working correctly. But here it will be argued that, like a "Kantian-style" explanation of success, the experimentalist's regress cannot successfully explain the *way* in which our theories have been confirmed.

The "experimentalist's regress" arises in the following way. Suppose a scientist wishes to test the correctness of, for example, the law $F = ma$. One way of testing this law would be to use apparatus to ascertain the magnitude of the force F acting on an object, the mass m of the object and the acceleration a of the object. It would then be a matter of elementary arithmetic to determine whether the obtained values were consistent with the equation $F = ma$. But, of course, if the results of these measurements are to either confirm or disconfirm $F = ma$ it must first be determined whether or not the apparatus used is working correctly. But how might we determine this? Very plausibly, any method we use to determine whether or not it is working correctly will at some point rely on the equation $F = ma$. But if so, we should hardly be surprised, or impressed, if when we use the apparatus to test $F = ma$ we get a confirmatory result. More generally, the experimentalist's regress raises the possibility that the "predictive success" of our theories may simply be a by-product of the fact that in determining whether our apparatus is working correctly, we use the very theories that the apparatus is intended to test.

However, the considerations given above show that the experimentalist's regress will sometimes fail to explain the *ways* in which our theories have been confirmed. Consider again, for example, the prediction of the planet Neptune. It is, perhaps, possible that if Galle had failed to find the predicted planet, he might have concluded that his telescope was not working correctly, or some other factor was present preventing him from detecting the presence of the planet.[16] But even if this is so, it provides no explanation at all of why he *did* find the planet when he looked in the spot suggested by Le Verrier. In this case, to explain why a spot *was* observed, we need to appeal to something more than just the way we check that our apparatus is working correctly. The same can be said about many other cases of novel predictive success. Perhaps if the first atomic bomb had failed to explode, the scientists involved would have concluded something had gone wrong with the mechanism, rather than that relativity or their theories of particle physics were wrong. But that does not explain why the bomb did explode.

There are, perhaps, some cases in which apparent predictive successes of our theories can be explained by "the experimentalist's regress", but it is also plain that there are cases in which it does not explain the way in which our theories led us to predict things which did happen.

SUMMARY OF RESULTS SO FAR

In this chapter we have considered three varieties of the success of science. These are (a) the ability of science to successfully lead us to novel predictive success, (b) the ability of science to give us knowledge of parts of the world that were not accessible at the time theories about them were first advanced, and (c) the ability of some *a priori* plausible theories to receive more subsequent confirmation than their experientially equivalent rivals. We have considered a wide range of possible explanations of these phenomena, including the suggestions that these successes are no more than we could reasonably expect by chance, that scientific realism could explain the phenomena, that they can be given "meta-inductive" or "evolutionary" explanations, that the "logic of invention" could explain them, that they could be explained by the hypothesis that the universe itself is simple, that sciences such as psychology or cognitive science might be able to provide the explanation, that a "Kantian-style" explanation might be given, or that they can be accounted for by the "experimentalist's regress". It has been argued that none of these approaches are satisfactory.

These negative conclusions naturally give rise to the question "What would count as a satisfactory explanation of the phenomena?" It is to this question that we now turn.

CRITERIA OF ADEQUACY FOR AN EXPLANATION

Earlier in this chapter we noted that the hypothesis of scientific realism by itself did not provide us with a satisfactory explanation of any of the phenomena. Whether or not the novel success of some scientific theories provides us with a cogent argument for the *truth* of scientific realism, the hypothesis of realism by itself does not *satisfactorily explain* the phenomena with which we are here concerned, because it does not explain how we have managed to *hit upon* theories that are true or close to the truth. Moreover, if we have managed to hit upon (approximately) true theories, that fact would stand in need of explanation at least as much as novel success itself. So we may state the first criterion of adequacy for an explanation of the phenomena as:

> *Criterion 1* A satisfactory explanation of the phenomena must provide us with an explanation of how *we have managed to hit upon* theories that (i) have novel predictive success, (ii) give us knowledge of hitherto inaccessible parts of reality and (iii) are empirically successful, despite being initially advanced on *a priori* grounds.

As we have already noted, we cannot tell in advance whether a theory is going to have novel predictive success, since we cannot see into the future. Neither can we tell in advance whether a theory tells us the truth about some part of reality that was not accessible to us at the time the theory was advanced. For example, neither Newton nor Adams nor Le Verrier knew of the existence of Neptune when putting forward their claims. Neither can we know in advance whether some *a prioristic* theory will turn out to be empirically successful. On the face of it, we use features such as *simplicity, plausibility, lack of ad hocness* and so on to identify in advance those theories that we believe will turn out to enjoy novel predictive success, give us knowledge of inaccessible parts of the world and be successful although *a prioristic*. Although it seems reasonably plausible to suggest that it is properties such as simplicity and plausibility that we use to identify in advance those theories we think will turn out to be successful, it would take detailed investigations into the history of science to say just what the properties we use actually are. Here we will simply call that property, or properties, that we use "property M".

At this stage of our enquiry, we will remain completely agnostic concerning the actual identity of property M. We will remain agnostic as to whether M is a single property or perhaps some (possibly highly disjunctive) collection of properties. We will also remain agnostic on the question of whether it is always the same property or properties that are used to identify theories that will turn out, in the various ways, to have novel success, or whether different properties are used on different occasions or historical periods. Finally, we will remain agnostic on the question of whether M is a single-place predicate or a many-place relational term. For example, we will remain agnostic on the question of whether a theory has M merely by virtue of having certain internal structural properties or whether it has M by virtue of bearing certain relations to other things.

Although we will remain agnostic concerning the actual identity of M, it is a consequence of criterion 1 that there are several features that M *must* have if it is to furnish us with a satisfactory explanation of the phenomena. First of all, M must be an *accessible* property of theories. Recall that we cannot perceive "directly" whether a theory will, in the future, enjoy novel predictive success or give us knowledge of less accessible regions or, if *a priori*, be empirically successful. We use the property M to identify those theories which, we think, will be successful in those ways. Plainly, if M is to play this role, it must be *possible for us to tell* whether, in fact, a theory has property M, and more specifically, it must be *easier* to tell if it has M than that it will be successful in the ways described. We may therefore state a corollary of criterion 1, which we call the *accessibility requirement*:

> *Criterion 1a* Property M must be *accessible*; more specifically, it must be *more accessible* than the forms of success of which it is taken to be an indicator.

There is another corollary of criterion 1 that is worth stating explicitly. If we are to explain why it is that scientists have hit upon theories with property M, it is not enough that property M be accessible, *there must also have been some reason why we preferred theories with property M rather than those theories with some other accessible properties.* Property M is, after all, undeniably going to be only one among a very large number of properties of theories that are easily accessible. The fact that a theory was invented on a Tuesday, or that it is stated with an even number of symbols, or that it contains an "=" sign, are all highly accessible properties of theories. But, at least on the face of it, there is no reason why scientists would be drawn towards, or prefer, or single out for testing, theories with those particular accessible properties. An explanation of the success of our methods must explain *why* it is that scientists prefer theories with *M* rather than one of the very many other highly accessible properties of theories. This leads us to our second corollary of criterion 1, which we will call the *explicability requirement*:

> *Criterion 1b* Property M must be such that we can explain why we have preferred theories with M, rather than any one of the many other highly accessible properties of theories.

The second criterion of adequacy for an explanation of the phenomena is that it must not restrict itself *merely* to an account of how we have managed to hit upon theories with M. Earlier in this chapter, we considered the suggestion that perhaps the "logic of invention", or perhaps sciences such as psychology or cognitive science, might be the appropriate disciplines to look to for an explanation of the phenomena. But we also noted that such disciplines would not provide us with a *complete* explanation. For example, cognitive science or psychology might give us an understanding of how the human brain or mind was able to discern simple patterns in some bewilderingly complex collection of data, *but such an account, by itself, would not explain why the theory or law stating that simple pattern went on to enjoy subsequent novel success*, or why it turned out to give us *knowledge* of parts of reality not accessible at the time that pattern was first perceived. Similarly, a program which enabled a computer to find a simple law in some body of data would not, in itself, explain why that law went on to enjoy subsequent novel predictive success, or why it gave us knowledge of less accessible realms. We may therefore state the second criterion of adequacy for an explanation of the phenomena as:

> *Criterion 2* An account which *merely* explains *how* we have managed to hit upon theories with property M would not be satisfactory; the account must also explain why it is that theories with property M (tend to) enjoy the forms of success exemplified in the three phenomena.

As we have already emphasized, criteria 1 and 2 would still not be sufficient to ensure a satisfactory explanation of the phenomena. It may be that we have one account of why we prefer theories with M and another, quite independent, account of why theories with M are successful. These two accounts together may not constitute a satisfactory explanation of the phenomena if they leave it as just a fortunate and highly improbable fluke that the type of theory that we prefer just happens to also be the type of theory that tends to be predictively successful. We have considered one example of a suggestion that would be unsatisfactory in this respect: the suggestion that we prefer theories that are simple because they are easier to understand and work with, but those theories are successful because the universe, fortunately, also happens to be simple. We also noted that this fortunate agreement between our preferences and the way the universe is would be *a priori* highly improbable. An explanation of the phenomena which ultimately relied upon such a highly improbable fluke would not be satisfactory. And so we may state the third criterion of adequacy for an explanation as:

> *Criterion 3* Any satisfactory explanation of the phenomena must not leave it merely as a fortunate fluke that the type of theory that we have preferred *also* happens to be the very same type of theory as that which (tends to) enjoy the forms of success exemplified in the phenomena.

There is one more criterion of adequacy that must be met by any explanation of the phenomena. The phenomena with which we are concerned are actual phenomena in the history of science. There are cases, in the history of science, of theories that have enjoyed novel predictive success, or that have given us knowledge of hitherto inaccessible realms, or that have turned out to be predictively successful despite their *a prioristic* nature. We will call these cases the *historical examples* of the phenomena, and it is *these* examples that we want explained. We can therefore state the fourth criterion of adequacy for an explanation as:

Criterion 4 Any satisfactory explanation must be able to account for the actual *historical examples* of the phenomena, in a way that meets the other three criteria of adequacy.

It is important to state criterion 4, because, as we will later see, it might seem rather doubtful whether the most promising way of meeting the first three criteria would also be able to meet criterion 4.

It has been argued that these four criteria are *necessary* for a satisfactory explanation of the phenomena, but I also think that together they are sufficient. Clearly, our overall aim is to explain actual examples of success, that is to produce an explanation that meets criterion 4. But any explanation of actual examples of success that also meet the first three criteria would seem to be a satisfactory explanation. And I think it is fair to say that it is criterion 3 that plays the pivotal role here: if it is not merely a fluke that the type of theory we have preferred also happens to be the type of theory that enjoys the forms of success, then we will be able to say that a theory is successful *because* it is of a type that we prefer, and so we will have an explanation of success. The other criteria state conditions that must be met if any putative explanation is to satisfactorily meet criterion 3.

We begin the task of developing a theory that meets these criteria of adequacy in the next chapter.

CHAPTER 3

A defeasible *a priori* justification of induction

We are seeking an explanation for certain types of scientific success. If we refer to the class of relevantly successful theories as S, the explanation of their success can be represented as:

> Explanans: The members of S possessed certain properties M, where those properties M were known to scientists prior to the subsequent success of the theories.
>
> ---
>
> Explanandum: The theories in S went on to enjoy subsequent surprising success.

What is the nature of the explanatory or inferential link that takes us from the explanans to the explanandum? In the previous chapter it was argued that it seemed unlikely that *empirical science* could provide us with a satisfactory explanation. On the face of it, the only other way of generating an explanation would appear to be by establishing some sort of *a prioristic* link between the explanans and the explanandum. That is, what is required is some sort of *a prioristic* argument establishing that *if* the members of S had certain properties known prior to their success, *then* the members of S would be, or at least would have an increased probability of being, subsequently successful.

An *a prioristic* argument of this sort is (one form of) what is standardly referred to as justification of induction. But the problem of justifying induction is one of the main problems of epistemology. It is a problem that many philosophers are perhaps inclined to put in the "too hard basket". It might be felt that if we must produce (one form of) a justification of induction to explain the phenomena with which we are here concerned, we have reached an impasse and can go no further. However, I believe this conclusion is too pessimistic. In particular, in this chapter it will be argued that some recent work by Laurence BonJour (1998; 2005: 98–105) on the justification of induction enables us to move forward on this problem and opens a way by which we can develop a satisfactory explanation of the phenomena. BonJour's approach resembles in some respects earlier work by David Stove (1986).

THE AIM OF THIS CHAPTER

It is useful to distinguish between the following two questions:

1. Is the approach to justifying induction, or ampliative inference in general, as developed by BonJour, defensible on general philosophical grounds?
2. What *specific* property, or properties, of theories is required to explain the phenomena with which we are here concerned, and can a preference for theories with these properties be given a justification of the sort offered by BonJour?

The two questions are, of course, distinct. The first question is perhaps most naturally seen as a question in general epistemology; the second is more appropriately located within philosophy of science or methodology. Both questions are large and difficult. We concern ourselves with the first question in this chapter; we begin addressing the second in the next.

THE PROBLEM OF INDUCTION

Let us begin by reminding ourselves of the traditional problem of induction, and why it has been thought to be so intractable. The traditional Humean "problem of induction" can be expressed in the following argument:

> Any justification of induction must, surely, be either *a priori* or *a posteriori*. We know *a posteriori* claims to be true by experience. An *a posteriori* claim may be either a bare report of experience or a claim arrived at from a bare report of experience by either a deductive or an ampliative inference. A bare report of experience cannot be sufficient to justify induction, and neither can a claim deductively inferred from such a report, since it cannot assert anything more than is implicitly asserted in the report of experience. A claim inferred from a report of experience by an inductive inference would beg the question.[1] Hence, induction cannot be justified *a posteriori*. But the only things we can know *a priori* are analytic truths which, being devoid of content, can never be sufficient to justify induction. So induction cannot be justified *a priori* either. Since induction cannot be justified either *a priori* or *a posteriori*, it cannot be justified.

A POSSIBLE STRATEGY FOR REPLYING TO HUME

One assumption in the above argument that many philosophers have questioned is that the only things we can know *a priori* are analytic truths. It is, of course, highly controversial whether this assumption is correct. As we observed in Chapter 1, it is useful to distinguish between the claims "P is known *a priori* to be true *with certainty*" and "P is known *a priori* to be true". Ignoring Gettier-type cases, we can say that belief P is *known* iff P is believed, P is in fact true, and it is rational, or justified, to believe P. So it is clearly possible for P to be known to be true even though P might not be known *with certainty*. This will be the case if that which justifies P, or which makes belief in P rational, fails to establish the truth of P with certainty. And this will surely be the case with most of the claims – certainly all of the empirical claims – we would ordinarily say we *know*. Moreover, at least on the face of it, there might also be claims that are advanced *a priori* and which are at least to some degree reasonable to believe, but which are not known with certainty.[2] Now, we can say (again ignoring Gettier examples) that P is known *a priori* iff the following conditions are met:

Condition (i)   P is in fact true;
Condition (ii)   P is justified, or rational to believe; and
Condition (iii)   Our rational or justified belief in P is not based on observation or experience.

But note that there is no requirement here that the justification of P, or that which makes P rational to believe, also establishes the truth of P with certainty, or makes it "epistemically necessary". So the standard definition of "*a priori*" leaves it as a possibility that there is a class of beliefs that, although known *a priori*, are *not* known *with certainty*. Moreover, it seems to be possible to give examples of such beliefs. The following conditional seems to be an example of such a belief:

If the probability of event E occurring is 1, *then* event E will occur.   (1)

Note that (1) is neither analytic nor necessary.[3] If an infinite number of fair coins are tossed, then the probability of at least one head coming up is 1. But it is *possible* for only tails to come up: so (1) is not (metaphysically) necessary.[4] It also could turn out to be false: it is "defeasible". And so we can also say that it does not follow from the very meanings of the words in (1) that it is true, and so it is not analytically true. Hence (1) is both contingent and synthetic. But (1) surely counts as something that is rational to believe. The extent to which belief in (1) is rational or justified is surely at least as great as that for typical empirical claims, such as "Everest is the tallest mountain in the world", which is something we would ordinarily have no hesitation in classifying as something we know. So, (1) clearly satisfies condition (ii) for something to be *a priori* knowledge. Moreover, it seems that the only experience necessary for it to be rational to believe (1) is that which is necessary to understand the meanings of the terms in it. No additional empirical knowledge is needed to make it rational to accept (1). So the rationality of our belief in (1) meets the third condition, that it not be based on experience. Does (1) meet the first condition of actually being true? Plainly, if we grant that it meets the *second* condition of it being rational to believe (1), then it is also rational to believe it is true. So it is *rational to believe* (1) is true. Of course, we do not have a proof of the claim that (1) is true, but in this respect (1) does not differ from very many empirical claims which we would say we know. The fact that we do not know with certainty that (1) is true does not show we do not know it all, merely that the claim "We know (1)" itself has a certain degree of riskiness. But, again, in this respect it does not differ from many of the other things we would say we know. I conclude that (1) is an example of synthetic, contingent but *a priori* knowledge.

If we are prepared to grant that (1) is an example of synthetic, *a priori* knowledge, there seems to be little reason to not grant the same for a conditional such as:

> *If* the probability of an event E occurring is 0.9999, then event E will occur. (2)

Of course, as the probability of E becomes less, we will become less sure that the corresponding conditional should count as knowledge: for example, would we say that the conditional was something we knew if the probability of E was 0.8? But this vagueness about the application of "know" can also be found with ordinary empirical statements: sometimes I might not be sure whether I ought to say that I *know* that the door is locked.

It is worth noting that although (1) and (2) are both synthetic *a priori*, neither seems to be particularly epistemologically dubious or mysterious. I think it is fair to say that the synthetic *a priori* often is seen to be epistemologically mysterious because it appears to be offering us "something for nothing": it claims to be able to produce real, factual knowledge without any input from experience, an epistemological "free lunch". While it is true that *some* claims to synthetic *a priori* knowledge have this characteristic, neither (1) nor (2) need to be like this. Consider (2). We do not get a "free lunch" with (2), provided that the degree of confidence with which (2) itself is asserted does not exceed 0.9999. More generally, we can in the following way bring out why at least some synthetic *a priori* claims can be reasonable to believe without thereby claiming we are epistemologically getting something for nothing.

The conditional sentence:

> If E will occur, then E will occur. (3)

is, of course, an empty tautology. It conveys no factual information. But it is, we may surely say, certain and incorrigible. Our degree of confidence in the truth of (3) is as high as possible. If we use "Cr" to denote epistemic probability, we may surely say:

> Cr(If E will occur, then E will occur) = 1. (4)

But let us now consider:

If Pr(E) = 0.9, then E will occur. (5)

Unlike (3), (5) does have some empirical content. It is a *synthetic* statement. But (5), unlike (3), is not certain. It is to a degree *risky*; it might be false. Although (5) is not certain, it is surely something reasonable to believe to a certain degree. Moreover, we seem to be able to specify how reasonable it is to believe (5). It seems right to say:

Cr(If Pr(E) = 0.9, then E will occur) = 0.9. (6)

In moving from (3) to (5) we have moved from a tautologous statement devoid of content to a synthetic statement that does have some content. But we have also moved from a statement that has the highest possible degree of epistemic probability to one with a somewhat lower epistemic probability. We have, as it were, "paid for" the increase in content with a reduction in epistemic probability. So even though (5) is a synthetic statement, provided it is believed with a degree of confidence of no more than 0.9, we would not seem to be "getting something for nothing" in believing it *a priori*. More, generally a conditional statement of the form "If Pr(E) = n, then E" can be *a priori* reasonable to believe, provided the degree of belief is constrained by:

Cr(If Pr(E) = n, then E) $\leq n$. (7)

Here is another example of a synthetic statement which we plausibly know *a priori*:[5]

> *If* a coin has been tossed a million times, and it has come up heads (8)
> every time, *then* the coin is not fair, i.e. there is a greater propensity
> for heads to come up than there is for tails.

Conditional (8) is clearly a synthetic statement, since the antecedent clause talks only about the outcomes of specific tosses, while the consequent asserts the existence of a propensity. Although we do not know with certainty that (8) is true, it is surely rational to believe it is true, and it would appear that all we need to see that it is rational is a knowledge of the meanings of the words in it. So (8) satisfies the conditions for being something we know *a priori*, even though it is synthetic.

In summary, it seems that there are some synthetic truths we can know, or at least have good reason to believe, *a priori*. Although the examples we have considered are not known with certainty, they still (plausibly) constitute knowledge in the sense of true, rational belief. In what follows, examples such as these will be referred to as "soft" or "merely rational" *a priori* knowledge, rather than as "hard" or "certain" *a priori* knowledge. It is worth noting that these examples of soft *a priori* knowledge are *a priori* in the perfectly standard sense that the only experience necessary to know they are true is that which is necessary to understand the meanings of the terms in them.

If we admit a category of synthetic, albeit soft or defeasible *a priori* knowledge, it might be possible to reply to Hume, since one of the assumptions of the Humean argument is that the only things we can know *a priori* are analytic truths.

BONJOUR'S ATTEMPT TO JUSTIFY INDUCTION

Laurence BonJour (1998: 208–16) has recently offered a justification of induction. He asserts that the following statement is knowable *a priori*:

> If all of the (sufficiently large number of) A that have been observed   I
> so far have been B, it is unlikely that this should be due to chance.

From (I) it is surely an *a priori* reasonable inference to:

> If all of the (sufficiently large number of) A that have been observed   II
> so far have been B, it is likely that this is not due to chance.

But now, to say that some condition is *not* due to chance is, very plausibly, to imply that there is a tendency or propensity for that condition to obtain. So it is an *a priori* reasonable, albeit defeasible, inference to:

> If all the (sufficiently large number of) A that have been observed   III
> so far have been B, it is likely that there exists a law-like tendency
> (or propensity) for A to be B.

If there exists a law-like tendency or *propensity* for A to be B, it is surely an *a priori reasonable*, although obviously defeasible, inference to the conclusion that all A are B. Therefore, BonJour says, we may conclude:

> If all the (sufficiently large number of) A that have been observed   IV
> so far have been B, then it is likely that all A are B.

(IV) is what we want from a justification of induction. But BonJour has argued that (IV) can be established *a priori*. Therefore, it seems, we have an *a priori*, but defeasible, justification of induction.[6] Note that is not claimed that the conditional "If (I), then (IV)" is *a priori* certain, only that it has *a priori* plausibility. In the terminology used here, it is, perhaps, an example of "soft" *a priori* knowledge.

There is a feature of this approach that is worth emphasizing: it attempts to justify the likely truth of a generalization by arguing for the existence of a law-like tendency or propensity. The existence of a *propensity* comes first, the justification of the inductive generalization comes second, on this view. This point will become important later when considering induction as an inference used in actual science.

The aim of this chapter is to defend an argument broadly along the lines developed by BonJour. However, it is worth noting at the outset that the argument defended here leads to a conclusion a little weaker than that advanced by BonJour. BonJour defends the conclusion "If all observed A so far have been B, it is likely that all A are B." If "likely" is taken to mean something like "has a probability of over 0.5", then the conclusion defended here is somewhat weaker than that. Here it will only be argued that "all A are B" is more likely than any other proportion. To this extent, therefore, the position defended here perhaps falls a little short of what we usually want from a justification of induction.

EVALUATION OF BONJOUR'S ARGUMENT

BonJour is well aware that objections can be raised against this (initial statement of) his argument. Some people might be inclined to reject his initial claim that I is *a priori*. BonJour himself does not say much in defence of the *a prioricity* of (I). He asserts that the only reason why someone would be inclined to deny it would come from doubts about the notion of *a prioricity* itself (1998: 208). And in this BonJour is, I think, on strong ground. Surely no *empirical* evidence is required for it to be rational to believe (I). So the claim "We can know *a priori* that it is rational to believe (I)" is very plausibly true. But from this it trivially follows that "It is rational to believe (I)" is also *a priori*. It is therefore very plausible to assert that (I) is something we know *a priori*.

The inference from (I) to (II) on the face of it seems plausible, although we will later see that, *under some but not all conditions*, there can be grounds for doubting it. BonJour recognizes, however, that the inference from (II) to (III) might be questioned (1998: 211). It might be that the propensity that exists is not for *all* A to be B, but just for those A which also have property C to be B. Here BonJour says it is important to distinguish between two very different types of case:

1. those cases in which possession of the additional property C is causally relevant to the possession of property B; and
2. those cases in which it is *not* causally relevant.

We will consider cases of type 1 first. It seems to be quite clear that cases of this type might actually exist. Suppose, for example, that whenever a particular remote tribe has been observed by anthropologists or tourists, it has always behaved in a bizarre and colourful manner.[7] In such a situation, we might not be justified in drawing the conclusion that the tribe behaves in a bizarre and colourful manner *all* the time. Perhaps they only behave that way when anthropologists or tourists are around. And, of course, it may be that the presence of the observers is caus-ally connected with their behaviour: perhaps they behave in the colourful way to, for example, "attract the tourist dollar". Here we would *not* be justified in making the inference from "Whenever this tribe has been observed, it has behaved in a bizarre and colourful manner" to the conclusion: "At all points in time, this tribe behaves in a bizarre and colourful manner". BonJour admits that this type of thing is a possibility, but holds that there is at least one version of the problem of induc-tion to which his account still offers us a solution. Suppose that we have somehow established that all observed A are B *and* that those observed A have property B *independently* of their having been observed by us. Even in such a situation, there is the problem of what, if anything, entitles us to draw the conclusion that *all* A are B. And it is *this* problem that BonJour claims his account can solve.

We can briefly note here that there is another possible way we could respond to this type of situation. Let us grant that it is a possibility that the fact that the observed A have been B is due, at least in part, to our presence as observers. Even so, BonJour's argument, it seems, still justifies us in drawing the conclusion that if *we observe* A in the future, they will be B. This falls short of justifying the claim that *all* A are B, but it nevertheless justifies *expectations about our own future observa-tions*. And so we can say that there is another form of the problem of induction which BonJour's argument enables us to solve. This point will become important later, when we consider the explanation of the phenomena with which we are here concerned. The crucial point is that the phenomena with which we are here con-cerned are phenomena that have been *observed by us*. Therefore, even if BonJour's argument only entitles us to draw the conclusion that there is a propensity for A observed by us to be B, we may still have a probabilistic explanation of *why we have observed* the phenomena with which we are here concerned.

"SHARPENING UP" OF BONJOUR'S ARGUMENT

Even if we grant that there are satisfactory ways of dealing with the objection that something might *appear* to conform to a regularity simply *because* it is observed by us, other difficulties remain. Suppose that, in addition to being black, all the crows we have observed also have some other property. This might be, for example, the property of being located *near* the observer, where this proximity to the observer does not cause the crows to be black. Let us assume, for example, that the observers of the crows are all located in Australia. So, instead of drawing the conclusion that there is a propensity for *all* crows to be black, we could, perhaps, draw the more limited and specific conclusion that there is a propensity for the crows in Australia to black. But then, if *that* is the propensity that exists, what, if anything, entitles us to draw the conclusion "All crows are black", rather than the more limited assertion "All crows in Australia are black"? Alternatively, perhaps it is not the case that there is a propensity for all crows to be black: perhaps there is a propensity for 99.999999 per cent of crows to be black. If this is the case, then maybe there are only three or four non-black crows in the world. In such a situation, we would statistically expect all the crows in Australia to be black, even though it is false that there is a propensity for all crows to be black. So it seems we cannot conclude that it is *a priori* likely that if all observed crows have been black, then there is a propensity for *all* crows to be black: there are a number of other propensities the existence of which would seem to be reasonably likely in the face of the fact that all the observed crows were black. Perhaps discovering that all the crows in Australia were black would show it to be unlikely that, say, there was a propensity for only 50 per cent of crows generally to be black. But showing all crows in Australia to be black would not show it to be unlikely that there was, say, a propensity for 99.9999999 per cent of crows to be black. Hence showing that all crows in Australia are black would not show it to be unlikely that "All crows are black" is false.

So, as it stands, BonJour's argument will not quite do. Nevertheless, I believe there is a way of modifying his argument so that it overcomes this difficulty. Let suppose that it is indeed the case that there exists a propensity for the crows *in Australia* to be black, but there is no propensity for crows generally to be black. Suppose that elsewhere they have, let us say, a propensity to be green. We live, as it were, on an island of black-tending crows in an ocean of green-tending crows. But in such a world, one thing that would be rather improbable would be our *location*. We would be located in a very unusual part of the universe. So if we say that we do indeed live on an island of black-tending crows surrounded by an ocean of green-tending crows, we are attributing to ourselves a highly unlikely location. Hence saying that we live on such an island of black-tending crows means that we have to make an assertion that *a priori* seems to have a very low probability of being true.

Of course, saying that we live on a small island of black-tending crows in an ocean of green-tending crows is not the only way of attributing to ourselves a rather unlikely location. Suppose, again, we notice that all the crows in our immediate environment are black. We could assert that in one half of the universe, there is a propensity for crows to be black, while in the other half there is a propensity for them to be non-black, and we just happen to live in the half in which there is a propensity for crows to be black. Pretty clearly, then, the *a priori* probability of us living in that half will be 0.5. Or again, we notice that all the crows in our region are black, and we assert that in 99.99 per cent of the regions of the universe crows have a propensity to be black, while in the remaining 0.01 per cent of the universe they have a propensity to be non-black. In this case, we are asserting that the probability of us having the location we do is 0.9999. *But* if we say that we live in a world in which *all* crows have a propensity to be black, and we say we live in a part of the world that has only black-tending crows, the probability of us having a location of this type will be 1.

Let us assume that we have established by observation that we live in a part of the world in which crows have a propensity to be black. What is the rest of the world like? There are many ways the world might be, four of which are as follows:

World A: We live in a part of the world in which there are only black-tending crows *and* the world in which we live consists of a small island of black-tending crows surrounded by non-black crows.

World B: We live in a part of the world in which all the crows have a propensity to be black *and* the world in which we live is divided into two halves: one half contains only black-tending crows, the other half containing only non-black-tending crows.

World C: We live in a part of the world in which all the crows have a propensity to be black *and* the world in which we live is divided into two parts: 99.99 per cent of the world consists of black-tending crows, while the remainder consists of non-black-tending crows.

World D: We live in a region of the world in which all the crows have a propensity to be black and the world in which we live consists entirely of black-tending crows.

Now, if we assert that the world we live in is World A, we are assigning to ourselves a location in this world which is *a priori* highly improbable. Similarly, if we assert World B or World C, we are asserting that our location in the world has an *a priori* probability lower than 1. But if we assert World D, then our location in the world does indeed have an *a priori* probability of 1. Since asserting anything other than World D requires us to assert something less probable than does asserting World D, it follows that World D is to be preferred to the others.

In the above argument, the expression "region of the world" was taken to mean "spatial region of the world". But, of course, the argument given would still apply if it were taken to mean "temporal region" or "spatiotemporal region". Hence we may assert that if we live in a spatio-temporal part of the world in which all the crows have a propensity to be black, the hypothesis that is more likely to be true than any other hypothesis is that all crows in all regions of space and time have a propensity to be black.

So far we have been considering the possibility that the world is divided into two sections, with one section containing crows that have a propensity to be black and the other not having that propensity. But this leaves it unspecified just what it means to say a crow has a *propensity* to be black. A "propensity" is a "tendency", but how strong must this tendency be? The fact that a crow has a propensity to be black need not *ensure* it is black. That is, it may be that:

If crow A has a propensity to be black, then the probability that crow A will be black is n, where n may be less than 1.

Suppose all crows observed so far have been black. Then, if we accept the argument given above, it will be rational to accept that there is *some* propensity for crows to be black. But how "strong" is that propensity? Does possession of the propensity ensure the crow will be black, or does it just confer on the crow some probability, less than 1, of being black? It is easy to see that the type of argument used above can show that it is most likely that the propensity that actually exists confers on crows a probability of being black of exactly 1.

Let us suppose that the following statements are both true:

(A) There is a propensity for crows to be black, and possession of this propensity confers on crows a probability of being black of 0.5; and
(B) We have observed some finite body of crows, and all of them have been black.

Clearly, the probability of obtaining the observations mentioned in (B), given that (A) is true, is less than 1. In fact, if possession of the propensity confers on the crows only a 50 per cent chance of being black, the chances of *all* of some (reasonably large) body of crows being black is rather low. Therefore, asserting that both (A) and (B) are true requires us to assert that something has taken place that has a probability of less than 1. More generally, we may assert:

> If, for any n less than 1, there is a propensity for crows to be black (1)
> that confers on a crow a probability of being black of n, and if all of
> some finite body of crows have been observed to be black, then the
> probability of us obtaining these observational results is less than 1.

On the other hand, we may also assert:

> If there is a propensity for crows to be black that confers on crows (2)
> a probability of being black of 1, and if all of some finite body of
> crows have been observed to be black, then the probability of us
> obtaining this observational result is 1.

Consequently:

> If all of some finite body of crows have been observed to be black, (3)
> then saying that crows have a propensity to be black which confers
> on them a probability of being black that is anything less than 1
> makes our observations less probable than does saying they have
> a propensity to be black which confers on them a probability of
> being black of 1.

And from this it follows that:

> If all of some finite body of crows have been observed to be black, (4)
> then the hypothesis that there is a propensity for crows to be black
> which confers on them a probability of being black of 1 is more
> likely to be true than any hypothesis which confers on them a
> probability of being black of less than 1.

Consequently, putting together the results of this section, we may say that:

> If all of some finite set of crows have been observed to be black, (5)
> then the hypothesis "All crows have a propensity to be black, and
> that propensity confers on them a probability of being black of 1" is
> more likely to be true than any alternative hypothesis.

It might be objected against this argument that (5) does not follow. Perhaps the assertion that there is propensity that confers on crows a probability of being black of 1 is more likely than any other *specific assignation of probability*, but it does not follow that it is more like than any other hypothesis at all. For example, it does not follow that the hypothesis that the probability is 1 is more likely than the hypothesis that the probability has *some* value less than one. This objection is, of course, entirely correct. Nothing in what has been said entitles us to assert that it is *likely* that the probability is 1, or that it has a probability greater than 0.5. But still, the argument, if valid, shows that the hypothesis that the probability is 1 is more likely than any other specific probability.

Let us say that if all crows have a propensity to be black, and that propensity confers on crows a probability of being black of 1, then crows have a *full propensity* to be black. But now it is surely rational to assert:

> If all crows have a full propensity to be black, then all crows are (6)
> black.

And from (5) and (6) it follows that it is rational to assert:

> If all of some finite body of crows have been observed to be black, (7)
> then it is more likely that all crows are black than that any other
> proportion are black.

A general objection might be raised against the view advocated here. The fact that all the crows we have observed so far have been black *might* be explained by saying all crows are black. But it might also be explained in some other way. Another possibility might be that *not* all crows are black but, whenever we have been observing crows, we have been wearing an item of clothing, or carrying with us some item of food, or making some sound particularly attractive to *black* crows. The number of possible alternative ways of accounting for our observations seems limitless. So what entitles us to conclude that it is most likely that our observations are due to *all crows* being black?

Let us suppose there is something about us that causes us to observe only black crows, or for it to seem to us to be the case that crows are black, even though, as a matter of (observer-independent) fact, it is not the case that all crows are black. Here we may distinguish between two different types of case.

The factor about us that causes it to be the case that we perceive (i)
only black crows is *always* present; and

The factor about us that causes it to be the case that we perceive (ii)
only black crows is *only intermittently* present in us: it "comes and
goes".

We have, in fact, earlier mentioned the case of (i) in our discussion of BonJour. It
was there argued that, *for the purposes of this book*, it does not matter if (i) is true.
Our concern is to explain the novel predictive success of science. If there is a factor
in us causing it to be the case that we see black crows, and that factor is *always*
present, then we will be able to predict that we continue to see black crows, and
also explain why we do so. More generally, we will be able to explain why highly
independent theories, based on data *we* have obtained, lead to observable predictions that *we* find to be confirmed.

Let us now consider the case in which the factor influencing our observations is
present only intermittently. For definiteness, let us assume that whenever we have
seen black crows, we have been wearing an item of clothing that attracts the black
crows but repels the others. But for this to have been the case, then an event must
have occurred which is, intuitively, rather improbable, and in any case clearly has
a probability of occurring of less than one. If we are in this way to explain why
we have only observed black crows, it must be the case that all the times in which
crows of any sort were around were also times in which we were wearing the black-
crow-attracting clothing, as if crows had been around when we were not wearing
this specific clothing, and therefore crows of other colours had *not* been repelled,
then there is at least some chance we would have seen these other-coloured crows.
We must therefore say that a less than maximally probable event has occurred.
The circumstances in which we obtained our data must, that is, have been quite
specific: they must have been restricted to the times we were wearing the black-
crow-attracting clothing.

It is now clear how the case presently under consideration is like the type of case
examined in the previous section. It was there argued that if we deny all crows are
black, then our data must have been obtained from unlikely, or at least less than
maximally probable, locations. A less than maximally probable event has occurred
with respect the circumstances of our data collection. But if we say all crows are
black, we do not need to say that any such less than maximally probable event has
occurred. The same can be said about the present case. Here, the circumstances
in which our data were obtained must have been unlikely, or at least less than
maximally probable: they must have been obtained only when we were wearing the
clothing that attracted black crows. But if we say that all crows are black, then we
do not need to say that any such less than maximally probable event has occurred.
As in the previous case, the hypothesis that all crows are black is therefore to be
preferred.

This completes our "sharpening up" of BonJour's argument for induction.

CRITICISMS OF BONJOUR

In his article "BonJour's *a priori* Justification of Induction", Anthony Brueckner (2001: 1–10) offers a number of criticisms of BonJour's argument. Perhaps his main criticism concerns the "interpretation of probability". BonJour says it is *a priori* reasonable to believe that if all A observed so far have been B, then it is *probable* that this is not due merely to chance. But what does *probable* mean in this context? Brueckner plausibly claims it cannot refer to *frequencies*. More specifically, it surely cannot mean something like:

> In the *majority* of possible worlds in which all A observed up to some point in time have been B, it has not been due to chance in those worlds that they have been B.

It is just not very plausible that we have *a priori* knowledge of a claim like that. But if it is not a claim about frequencies, what is it? Brueckner considers the possibility that BonJour's claim is to be understood as being like the following claim:

> The following is knowable *a priori*: "Bill tossed a dime ten times" makes probable "Bill did not toss ten heads".

Brueckner is apparently willing to concede that there is some sense of "probable" on which this claim as at least *prima facie* plausible. However, he points out that actually the part of it that appears after the colon is knowable *only if* the additional information is given that Bill was tossing a *fair* coin; and we cannot avail ourselves of information of this sort for the purposes of justifying inductive inferences in real-world cases.

Brueckner is surely right that we need some information about, for example, the fairness of the coin if we are to reasonably claim to know the above assertion about Bill tossing the coin. *But the inference used by BonJour does not require information of this sort.* The inference used by BonJour can be represented as follows:

> It can be known *a priori* that "All observed A are B" makes likely "It is not due to chance that all observed A are B (and hence that there is a propensity for A to be B)".

This inference does not require us to make any assumptions about the probabilities of certain types of outcomes; it is rather an *inference to* such probabilities. This can perhaps be made clearer if we note that the type of claim made by BonJour is not like Brueckner's example of Bill tossing the coin given above, but is rather more like the following:

> It can be known *a priori* that "Bill tossed the coin ten times and it came up heads every time" makes likely "It is not due to chance that

heads came up every time (hence, there is a propensity for heads to come up, i.e. the coin is *not* fair)".

This does not require us to make any assumptions about the fairness of the coin; it rather warrants us in *drawing a conclusion* about its fairness or otherwise from the data. I conclude that Brueckner's objection fails.

Another set of criticisms of BonJour have been developed by John Meixner and Gary Fuller (2008: 227–9). The core of Meixner and Fuller's objection concerns a particular step in BonJour's reasoning. According to Meixner and Fuller, BonJour assumes that if a large number of A have been observed to be B, there are only two possibilities:

1. It is merely due to chance; or
2. There exists a propensity for A to be B.

But Meixner and Fuller point out that these do not exhaust all the possibilities. The other possibilities mentioned by Meixner and Fuller are:

- a "grue"-type hypothesis – that all A do indeed have *some* property but that property is not B but rather some grue-like alternative;
- that the regularity observed is not universal but limited in some way in time or space; or
- that the causal law responsible for the regularity is likewise limited in time or space.

The second two possibilities – that the regularity, or the causal law responsible for it, is limited in time and space – were discussed earlier in this chapter. "Grue"-type alternatives to conventional predicates are discussed in the next chapter.

THE RELATION BETWEEN THIS ARGUMENT FOR INDUCTION AND OUR ACTUAL INDUCTIVE PRACTICES

It is worth re-emphasizing that the approach advocated justifies induction via the notion of a propensity. As we will see, doing things this way has a number of advantages. One advantage is that this closely parallels the way we actually do seem to go about confirming scientific hypotheses. How do we actually go about confirming that, for example, all salt melts at 801°C? The first thing to note is that it only requires the existence of the observation of a few samples of salt that melt at 801°C to persuade us that *all* samples of salt will melt at that temperature. This is explicable on the account offered here. It is highly unlikely that it should just be due to chance that three or four randomly chosen samples of salt should all melt at *exactly* 801°C, and so we find it natural to reason that it is not due to chance, and hence that there is a propensity for salt to melt at this temperature. But if there is a propensity for salt to melt at this temperature, it is more likely than any other general hypothesis that all samples will do so.

There is another reason to believe that we arrive at universal generalizations via the notion of a propensity. Suppose we have observed ten black things and noted they have all been crows. We could draw the conclusion that all crows are black, but we could, it seems, also draw the conclusion that all black things are crows. Both generalizations have received an equal number of positive instances. But even if we had never encountered any black things before that were non-crows, I think we would be more inclined to draw the conclusion that all crows are black than to draw the conclusion that all black things are crows. And one possible reason for this is because, while we are quite prepared to accept that there may be a propensity for crows to be black, we feel it is much less likely that there could be a propensity for black things to be crows. The very notion of a propensity for black things to be crows seems bizarre.[8] We feel the blackness of a thing could not cause it to be a crow. If there is no such thing as a propensity for black things to be crows, the inference to the conclusion that all black things are crows is, on the account offered here, blocked. More specifically, the inference from (II) to (III) discussed earlier (page 44ff.) is blocked. The same argument can explain why we are inclined, after a few positive instances, to accept that all salt melts at 801°C, but are not inclined to accept that all things that melt at 801°C are salt: while we have little trouble accepting the idea of there being a propensity for salt to melt at 801°C, we regard it as unlikely, or indeed impossible, that there should be a propensity for things that melt at 801°C to be salt. So, in summary, there seems to be some evidence that in our actual inductive practices, we arrive at universal generalizations via the notion of a propensity. It is, of course, desirable for our purposes that the account of inductive inference advocated here mirror our actual inductive practices, since this account is meant to be an explicit statement of the implicit knowledge we have of those practices.

Another advantage associated with using the notion of a propensity is that it helps to explain the *explanatory* and *law-like* character of generalizations arrived by induction. If crows have a propensity to be black, then we can say that a particular object is black *because* it is a crow. We can also say, for example, that although there is no crow on that branch, if there had been it would have been black.

In considering how universal generalizations are confirmed, we can distinguish three broad types of case, which we will call "the positive case", "the neutral case" and "the negative case". Whether a case is positive, negative or neutral depends on how *quickly* we regard it as being confirmed by positive instances. In cases of the positive type, we very quickly become convinced that a universal generalization is true after just a few positive instances. Examples of the positive case are quite common. Some examples are "All salt melts at 801°C", "All silver conducts electricity", "All hydrogen is inflammable" and so on. In the *negative* type of case, even a very large number of positive instances would fail to convince us of the truth of the generalization. Examples of the negative type of case are also fairly common. Some are "All things that melt at 801°C are salt", "All black things are crows", "All mortals are Greeks" and so on. Cases of the *neutral* sort are much rarer. We are confronted with a neutral case when we encounter a "black box" of unknown origin and constitution. Suppose we note that whenever, in the past, we have touched the black box, it has hummed. We would, I think, fairly quickly become convinced of the truth of the generalization "Whenever we touch the black box, it will hum." We would certainly become convinced of its truth more quickly than we would in cases of the negative sort, although perhaps not so quickly as we would in cases of the positive sort.

That generalizations should be divided in to these three classes is easily explainable. In the negative case, our background beliefs tell us it is highly unlikely that there should be a propensity of the required sort, that is a propensity for things that are mortal to be Greek or for things that melt at 801°C to be salt. So the inference to the generalization is blocked.

In the neutral case, we do not have any prior background evidence that a propensity of the required sort is likely to exist or likely not to exist. Nevertheless, even in the neutral case, I think we are inclined to accept the universal generalization after just a fairly small number of positive instances. And this is in keeping with the view advocated here, according to which the universal generalization "All A are B" becomes more likely than any alternative even after just one positive instance.

In the *positive* case, our past experience tells us, for example, that samples of the same chemical substance are likely to have the same physical propensities, such as melting at a particular temperature or conducting electricity. So, in positive cases, there is, on the view advocated here, additional evidence for the propensity not present in the neutral case.

DOES THE ARGUMENT DEVELOPED HERE GIVE US ALL
THAT WE MIGHT WANT FROM A JUSTIFICATION OF
INDUCTION? WILL IT DO THE JOBS WE WANT DONE?

In this section it will be argued that, although there are two respects in which the account offered here falls short of being a justification of induction in this sense, it nevertheless comes at least pretty close to doing the jobs we want done here. In particular, it has many of the features a justification of induction must have if it is to explain the three phenomena described in the first chapter.

First, let us see how the account offered here falls short of having *all* the features we would ordinarily want from a justification of induction. It is usually thought that if a justification of induction is to be satisfactory, it must show that it is *likely* that all A are B. To say that something is *likely* surely entails that it is more likely than not, and hence that its probability is greater than 0.5. But we have plainly not shown this here. What has been argued is that if all observed A have been B, then the hypothesis that *all* A are B is *more likely than* any other claim of the form "*n*% of A are B", where *n* is some number less than 100. But this is clearly compatible with "All A are B" having a probability less than 0.5. It may have a probability of much less than 0.5; it is just that, if the view presented here is correct, all other hypotheses have an even lower probability. So the argument offered here does fall short of what we usually want from a justification of induction.

Nonetheless, the argument offered here may still be suitable for our purposes. The aim of this book is to offer an explanation of the success of certain scientific theories. And in order to explain those forms of success, it might not be necessary to show our theories are *likely*. It may be sufficient merely to show that having inductive support increases the probability that our theories will be empirically successful.

First, let us note that there are some models of explanation on which an explanans does not have to confer upon the explanandum a probability greater than 0.5. One such model is Wesley Salmon's (1971) statistical relevance account. On Salmon's view, for some factor F to explain an event E, it is sufficient for F to be statistically relevant to the occurrence of E, or for the occurrence of F to raise the probability of E. And it seems fairly likely that the argument for induction offered here might enable us to give explanations of that sort for the success of science. On the account offered here, if a generalization G can be arrived at by the straight rule from observed data, it is more likely to be true than other generalizations also compatible with the obtained data. It is, therefore, rendered more likely than any randomly chosen generalization also compatible with the obtained data. Consequently, the probability of the generalization is increased by its being derived by the straight rule from obtained data. Therefore, the probability of any sentence P derivable from the generalization G is also increased. But now, suppose that P is a sentence derivable from G, and that P is found by subsequent observations to be true. The fact that G was derivable by the straight rule from earlier observations therefore provides us with a probabilistic explanation of why P was subsequently found to be true. Hence the justification of induction offered here, if correct, might provide us with probabilistic explanations of predictive success. It may fall short of doing all we would like from a justification of induction, but it may still be sufficient to do the jobs we are concerned with in this book.[9]

There is another respect in which the argument of this chapter falls short of being a "full" justification of induction. A crucial step in the argument is that from "It is highly unlikely that it could be due to chance that all observed A are B" to "It is therefore likely that there is a propensity for A to be B". But, as we noted earlier, it is possible to object to this step. Perhaps the propensity is not for all A to be B, but for all *observed* A to be B. BonJour's response to this objection, we recall, was to assert that one centrally important traditional problem of induction was – assuming it to be an observer-independent fact that some finite body of A are B – to show that we are entitled to draw the more general conclusion that all A are B. He asserts that his justification is at least a solution to that problem. But we also noted that there is another possible response to this difficulty. Whether crows appearing black to us is due to their really being black, or whether it is due in some way to our presence as observers, the argument, if sound, shows that it is more likely that if *we observe* crows in the future, they will be found to be black than any other result. And, if we do find crows in the future to be black, the argument provides us with a probabilistic explanation of why we should have found this. More generally, whether the observations we obtain are due to us perceiving things as they really are, or whether they are somehow due to our presence as observers, the account given here would seem to be able to explain the success of inductive inferences in leading us to future *observations*.

Plainly the above considerations may enable us to explain the ability of our theories to successfully make observational predictions, but they would not appear to hold out much hope of explaining *novel* observational predictions, or the other phenomena described in the first chapter. Mere enumerative induction or the straight rule would neither of them seem likely to be able to explain the phenomena. But in later chapters it will be argued that the type of argument defended here can be extended to do these jobs.

CHAPTER 4

The independence of theory from data

In the previous chapter it was argued that a theory arrived at by the straight rule from some body of data is more likely to lead to true predictions than other theories also compatible with the same data. This gives us a reason to prefer theories arrived at by the straight rule. But if we are confronted with any finite body of data, there will generally be a number of theories capable of being derived from those data by the straight rule.

Perhaps the most obvious way (to philosophers, at least) of generating indefinitely many generalizations from the same body of data using the straight rule is by using "grue/bleen" type predicates. The grue/bleen problem is discussed later in this chapter. But even without grue/bleen type predicates, it is still possible to use the straight rule to generate a number of different hypotheses from the same body of data. The history of science contains many such cases. For example, it may well have been the case that in the fifteenth century both the theory of Copernicus and a version of the Ptolemaic theory could both account for all the data that had been obtained up until that time. It would therefore have been possible to use the straight rule to make both of the following two inferences:

- Inference A: All available data conform to (a particular version of) the Ptolemaic theory; therefore all data (including data that will obtained in the future) conform to that theory.
- Inference B: All available data conform to the Copernican theory; therefore all data (including data that will be obtained in the future) conform to that theory.

Inferences like these could also be made about, for example, the oxygen theory of combustion and the version of the phlogiston theory advocated by Henry Cavendish with respect to the empirical data available at around 1785. Or again, similar inferences could be made about the interpretations of the Michelson–Morley result due to Lorentz and Einstein. More generally, it very often occurs in science that all the data available at some time can be seen as conforming to two or more *different* theories. The straight rule, by itself, does not tell us which theory to select. We need something more than the straight rule if we are to account for the choices made by scientists. Moreover, since the theories just mentioned – the theory of Copernicus, the oxygen theory of combustion and Einstein's Special Theory of Relativity – all enjoyed novel predictive success, we need something more than the straight rule if we are to explain the phenomena with which we are here concerned.

SOME GENERAL FEATURES OF THE PROPERTY, OR PROPERTIES, OF THEORIES WE NEED TO EXPLAIN NOVEL SUCCESS

What *type* of property of theories, over and above inductive support, is likely to furnish us with an explanation of our phenomena? We will initially focus our discussion on phenomenon 1, the phenomenon of novel predictive success. A theory enjoys novel predictive success if it successfully predicts observational regularities different from any of those on the basis of which it was initially formulated. For example, if a theory T that was initially based on three observational regularities R_1, R_2 and R_3 successfully predicted instances of another observational regularity R_4, then theory T would (plausibly) have thereby had a novel predictive success. But since theories are always underdetermined by the data on which they are actually based, there will be many theories – T^*, T^{**}, T^{***} and so on – other than T, all of which would have been able to explain the three initial observational regularities. Most of these other possible theories will not successfully predict the novel observational regularity R_4. Yet somehow we managed to select, in advance, one of the theories (theory T) that *was* able to successfully predict the novel regularity. As was argued in Chapter 2, it was unlikely this was due merely to good luck. So, it seems, we must have been able to recognize in T some property it had which made it more likely to be successful than the other theories. We will continue to refer to this property as "Property M". Now, it seems pretty clear that, whatever M is precisely, it must surely be something rather similar to *simplicity*, since it is such a widely accepted belief that we do in fact use simplicity to choose among rival theories. If property M is not exactly the same as simplicity, it ought to be able to furnish us with an explanation of why it is so widely believed that simplicity is best.

THE INDEPENDENCE OF THEORY FROM DATA

One of the main theses defended in this book is that the notion which does *most* of the work in explaining the three phenomena with which we are here concerned is the notion of the *independence of theory from data*. It will be argued that theories with this property are likely to have subsequent empirical success and that this claim can be given a justification similar to the justification of induction given in the previous chapter. It will also be argued in Chapters 6–8 that many actual cases of theories that exemplify our three phenomena are theories that have high degree of independence from the data. The notion of the independence of theory from data plays the central role in this book. The purpose of this section is to introduce this notion.

It is worth emphasizing at the outset that it is *not* here being claimed that theories with a high degree of independence from the data are more likely to be true or close to the truth, where this is understood in "scientific realist" terms. No such claim is made. It is merely claimed that having a high degree of independence from the data gives a theory an increased likelihood of subsequent empirical success. An immediate advantage of such an approach is that it thereby allows, at least in principle, an explanation of theories that do not by our lights appear to be even close to the truth but which were, nonetheless, surprisingly successful.

Consider the following sequence of numbers:

$$2, 4, 6, 10, 14, 22, 26, 34, 38, 46, \ldots . \tag{S1}$$

It might be an exercise in a school test to find the rule that generates these numbers: we are probably all familiar with questions of this type from our childhood. But there are several rules that are capable of generating the numbers. One rule that will generate the numbers is the following:

> Start with "2", then write down "4", then write down "6", then write   (R1)
> down "10", then write down "14", …, then write down "46".

Clearly, R1 is no briefer than the sequence of numbers itself. There would certainly be no practical advantage to remembering R1 rather than the sequence of numbers itself. However, R1 still qualifies as a rule for generating the numbers, since it is a set of instructions which, if followed, would result in the sequence that we want. But there are, of course, also some simpler rules that would give us the sequence. Here is one:

> "Starting at "2", add the lowest even number twice, then add the   (R2)
> second lowest even number twice, then add 8 and 4 alternately.

Finally, here is a third rule for generating the sequence:

> Put in the *n*th position of the sequence the *n*th prime number   (R3)
> multiplied by 2.

We intuitively feel that R3 is the best of the three rules. But why do we feel that it is the best? Perhaps the standard answer to this question is "because R3 is the simplest". But I don't think this answer does full justice to why we feel that R3 is the *best*.

We get a bit closer to the truth, I think, if we say that R1 is particularly bad because it is so ad hoc. R3, by contrast, seems much less ad hoc. Indeed, it will be argued that it is precisely because R3 is less ad hoc than R1 that it is better. *The notion of independence that occupies the key role in this book is essentially an explication of the idea of lack of ad hocness.* But still we are left with the question "Why should less ad hoc theories be better?"

Let us contrast R3 with R2. There are what we would intuitively judge to be *two* distinct patterns described by R2. The first of these is: "starting at two, add the lowest even number twice, then add the second lowest even number twice", while the second pattern is "add 8s and 4s alternately". One reason why we are sceptical about the correctness of this compound "rule" is because we feel it *might just be due to chance* that the numbers happen to exemplify this pair of patterns. It might be, as it were, an unintended, and in that respect "accidental", consequence of the real rule the sequence is following that it also exemplifies these two other patterns for a while. Now let us consider R3. According to this rule, we get the numbers in the sequence just by doubling the prime numbers. One reason why we feel that this is the correct rule is because we feel that *it could not merely be due to chance that such a long sequence of numbers obeys this rule.* It might well be due to chance – in the sense of being an unintended consequence of the correct rule – that the sequence, for a while, conforms to the rule: "add 8s and 4s alternately". But it seems much less likely that it should just be a fluke that the whole sequence conforms to the rule "double the prime numbers". We are inclined to reason in the following way: "It couldn't be just a fluke that the numbers conform to this rule. Therefore it *isn't* just a fluke. Therefore, this must be the rule that the numbers are actually obeying."

The above is not intended as a rigorous justification of the correctness of R3. It is rather intended as an articulation or "spelling out" of the reasoning that we in fact use when come to the conclusion that R3 is more likely to be correct than R2.

R1 and R2 are "bad" theories, and R3 is a "good" theory. Let us try to say a little more clearly why it is that R1 and R2 are bad. The basic idea is that we do not find them convincing because it could just be due to chance that the numbers conform to these rules. But what is it about these rules that causes us to come to this conclusion? Both these rules have a high degree of ad hoc dependence on the sequence of numbers. R1 is, we may say, totally dependent on the sequence: it is in effect just a restatement of the sequence itself. For each number in the sequence, there is some corresponding phrase in R1 which is determined by the number it is required to generate: the first number in the sequence is "2" and the first phrase in R1 is "Write down number 2", and so on. Or, we can say, each phrase in R1 is produced ad hoc for each number it is meant to generate. In this sense, R1 has the maximum degree of *dependence* on the sequence of numbers it is meant to generate.

Although R2 does not exhibit the same degree of dependence as R1, it still has many features that are dependent on the sequence of numbers it is meant to generate. The first part of R2 says: "Starting at 2, add the lowest even number twice." It is dependent on the data that the starting point should be "2", it is also dependent on the data that we should then proceed to *add* numbers, and it is also dependent that the numbers added should be *even* numbers, and that they should be added exactly *twice*. And so on.

R3 is, by comparison, much less dependent on the data than either R1 or R2. Of course, R3 contains some features that are dependent: for example, R3 is dependent on the data that it refers to *prime* numbers, rather than numbers of some other kind, it is also dependent that the prime numbers are to be *doubled*, rather than multiplied by some other number. That R3 contains these features (prime numbers, doubling) is dependent on the sequence of numbers that needs to be generated by the rule. But R3 seems to have considerably fewer such dependent components than either R1 or R2. We will say that while R1 is highly dependent on the data and while R2 is somewhat dependent on the data, R3 achieves a higher degree of *independence* from the data.

The notion of *independence from the data* can be regarded as an explication of the idea *of lack of ad hocness*. This idea has two dimensions:

1. The *"post hoc" dimension*. A theory can be said to be ad hoc, or dependent on the data, only if it is postulated *after*, and in response to, the observed features of the data. This is a necessary, but not sufficient, condition for *ad hocness*.
2. The *complexity dimension*. A theory that is post hoc will also be more ad hoc if and only if, we may roughly say, it is more complex. Exactly how this complexity is to be measured will be discussed later in this chapter.

Note that conditions 1 and 2 are both necessary for ad hoc dependence. This means that a theory that is very complex, but not postulated post hoc in response to the data, will not qualify as dependent on the view advocated here. This will become important later in this chapter and in subsequent chapters.

We can now spell out in a little more detail why it is we feel that rules such as R3 are more likely to be correct than rules such as R1 and R2.

Let us focus our attention on the intermediate sequence, R2. We can say that there is a reasonable possibility that the sequence of numbers should conform to a rule as dependent on the data as R2 *by chance*. But what might it mean, in this context, to say that the data conforms to R2 *by chance*? One obvious sense is as follows: it is *a priori* rather unlikely that a purely random collection of numbers should conform to a rule as simple of the data as is R2. But there is also another sense in which we can say it may be just due to chance that the sequence conforms to R2. Consider again the sequence S1. It contains ten numbers. Now let us suppose that there is in fact some rule that is used to generate the numbers in this sequence, and that this rule has been used to generate not just the first ten numbers in the sequence, but the first one thousand. The second and final component of R2 says "add 8s and 4s alternately". R2 therefore leads us to expect that, once we get past the first ten numbers in the sequence, they will continue to increase by 8s and 4s alternately. So, if we were to see one thousand of the numbers in the sequence, R2 might turn out to be either confirmed or falsified. Now, one sense in which it may just be due to chance that S1 conforms to R2 is that it may only be the first ten numbers in the sequence that conform to R2, and it is just due to chance that we have, as our data to work with, the first ten numbers in the sequence. Perhaps, if instead of having numbers 1 to 10, we had numbers 20 to 30, R2 would have been falsified. So, the fact that S1 conforms to R2 may be due to the fact that S1 is a particular data-subset that just happens to conform to R2. This is another sense in which it might just be due to chance that the data conform to R1. We will refer to *this* sense in which the adequacy of R2 may just be due to chance as *chance conformity due to the contingent location of the actual data*. It is *this* sense in which numbers – or data of any kind – might "conform to a rule by chance" that will be important when it is argued that rules that are highly independent of the data are more likely to lead us to correct predictions about future data.

Now let us consider R3. The first ten numbers in S1 conform to R3. Let us assume that S1 is indefinitely long, but that at this stage we do not know what pattern, if any, S1 exhibits once we get past its first ten places. What are the chances that ten consecutive numbers, chosen at random from the indefinitely long sequence that is S1, should exemplify a rule as *independent of the data* as R3? Intuitively, we feel that the probability is very low. The probability of chance exemplification of this rule seems to be so low that we seem to be justified in drawing the conclusion that it is *not* merely due to chance that the numbers should exemplify such a rule. But now, just what is it precisely that we feel could not be due to chance? It is highly unlikely that the *particular portion* of S1 that we have – the first ten numbers in it – should just happen to be a section that exemplifies this rule. The chances of exemplification due to the contingent location of the data seem to be very low in this case. Since this section of the whole sequence exemplifies this rule, and it is highly unlikely that the portion we have selected at random should just happen to conform to it, it seems reasonable to conclude that the whole sequence probably conforms to this rule. This is, in a nutshell, the reason why rules that are more independent of the data are likely to subsequently prove to be empirically correct: it is less likely that it should be just due to the contingent location of the data that the data should exemplify a highly independent rule.

THE DISTINCTION BETWEEN DATA EXEMPLIFYING A RULE AND CONFORMING TO A LAW OF NATURE

So far, our discussion has been concerned with the sequence of numbers S1. We can imagine this sequence of numbers occurring in a school test in which students are asked to find the rule by which the numbers are generated or to complete the next few numbers in the sequence. In these cases, there is nothing that determines which answer is correct above and beyond the intentions or opinions of the examiner. But, of course, scientists are not concerned with cases such as these. Scientists are concerned with cases in which the data is generated by nature itself. For such cases, it is possible to draw a distinction between a body of data *exemplifying a rule* and it *conforming to a law of nature*. Consider again the sequence S1. Let us suppose, this time, that this sequence was obtained as a result of observations of nature. We could, for example, imagine the numbers to be the results of meter readings, or, perhaps, of observations of the population of some organism. This sequence of numbers will *exemplify* all three rules R1, R2 and R3. A body of data exemplifies a rule R iff that rule R can be used to correctly derive that particular body of data. We can tell, with certainty, whether or not a body of data exemplifies some rule: it will merely be a matter of mathematics or logic whether or not it does exemplify a particular rule. But it does not necessarily follow from the fact that some data exemplifies a rule that it is also conforming to a law of nature. A body of data *conforms to a law of nature* R iff the data have a propensity or law-like tendency to exemplify that rule. We can explicate this idea as follows. Suppose measurements of some property have been obtained for values $V_1, ..., V_n$, and it has been found that, for those values, the measurements obtained all exemplify R. To say that the data merely exemplify a rule does not carry with it any implication that the data would continue to exemplify R for values other than $V_1, ..., V_n$; but if it is said that the data conform to a law of nature, it *is* implied that values of the data other than those for $V_1, ..., V_n$ would exemplify R. Or again, if it is merely claimed that measurements of the values of some property exemplify a rule, then it is not implied that if that property had been measured on a different occasion, or under different conditions, then the obtained values would still exemplify the same rule, but if it is asserted that the data conform to a law of nature, it is implied that they would exemplify the same rule under the different conditions. More generally, to say that the data conform to a law of nature is to imply that the fact that the data conform to a particular rule is *not* due merely to the *contingent location of the data*, but that the data would have conformed to that law under conditions other than those under which it was actually obtained.

THE DEFINITION OF INDEPENDENCE OF THEORY FROM DATA

We are now in a position to give an (explicative) definition of the notion of the independence of theory from the data. Our aim in giving an explicative definition of independence, and the inverse relation of dependence, is to get clearer on the idea of the extent to which a theory has just been "cooked up" in an ad hoc way in response to the data. Let us suppose that the numbers in S1 are data or readings that have been obtained as a result of the observation of some part of nature, and that R1, R2 and R3 have been advanced as *explanations* of those data. Intuitively, R1 is the explanation with the greatest degree of ad hoc dependence on the data. It has as many distinct explanatory clauses (each one of which depends on the datum) it explains for its content as the data have. Note here that both dimensions of ad hoc dependence – the post hoc dimension and the complexity dimension

– are involved. Each distinct explanatory clause of R1 is postulated post hoc in response to the observed features of the sequence. So the post hoc dimension is involved. Each clause is a *dependent* explanatory component. But since there are many such clauses, the complexity dimension is also involved. The larger the number of dependent explanatory components, the greater the complexity, and so the higher the degree of ad hoc dependence. The degree of dependence of R1 is very high: the number of dependent explanatory components (DECs) in R1 is, we may very roughly[1] say, no fewer than the number of components of data that it explains. By comparison, R2 has fewer DECs than R2, even though the number of components of data that it explains is exactly the same. So the *ratio* of DECs to components of data in R2 is lower than it is in the case of R1. Finally, R3 has even fewer DECs than R2, even though it explains exactly the same data: it has an even lower ratio of DECs to components of data explained.

The *higher* the ratio of DECs of a theory to components of data explained by a theory, the more dependent a theory is on the data. We may therefore define *dependence* as follows:

$$\text{Dependence of T} = \frac{\text{Number of DECs of T}}{\text{Number of components of data explained by T}}$$

We may define the degree of *independence of T from the data* as the inverse of its degree of dependence:

$$\text{Independence of T} = \frac{1}{\text{Dependence of T}} = \frac{\text{Number of components of data explained by T}}{\text{Number of DECs of T}}$$

It is worth stressing that the degree of independence of a theory is not a measure of its simplicity *per se*, or even of its simplicity relative to the amount of data it explains. It is a measure of its lack of ad hoc dependence on the data. The independence of a theory is not lessened by its merely being complex, or having a large number of explanatory components. Its independence is, rather, lessened by its having a large number of *dependent* explanatory components, that is components postulated post hoc in order to explain observed features of the data. A theory with many explanatory components, but where those components have independent evidence in their favour and are not merely postulated post hoc, might actually have a high degree of independence.

The question of how we individuate or count DECs and components of data explained is considered below (page 86ff.). However, we can here make some broad, general remarks about the possible values that can be taken by the degree of independence of a theory. A theory will have a degree of independence of 1 iff it has as many DECs as components of data it explains. Generally, and as we will later see, in order to be *good*, a theory must have a degree of independence greater than one. It is nonetheless *possible* for a theory to have a degree of independence less than one. Plainly, this will occur if the theory has a larger number of DECs than components of data it explains. This situation might arise if a theory is advanced which contains a number of distinct components each one of which by itself would be capable of explaining the data. For example, suppose Jones arrives home and finds the front door open. He explains this by saying that his wife arrived home and left the door open, his son arrived home and left the door open and his daughter arrived home and left the door open. Here there are, we may (roughly and intuitively) say, three DECs used to explain a single component of data, and so Jones's theory has a degree of independence of one third. It is possible for a theory to have a degree of independence as low as zero.

There is no *upper* bound on the degree of independence that a theory may have, since there is no upper bound on the number of components of data that may be explained by a single DEC, or by some finite number of DECs.

Not all components of a theory need to be *dependent* components of theory. A component is *dependent* on the data if the reason for asserting that component lies in the data itself. But it is *possible* that the reason for asserting a particular theory might not lie within any data. For example, if a theory is known *a priori* to be true, then it will not have any degree of dependence on any data. Moreover, as we saw in Chapter 1, it seems as though there may be some cases of *a priori* theories that do explain data. If so, such theories would have no dependence at all on the data they explain so would have the highest possible degree of independence from the data.

In summary, the value of independence of theory from data can range from zero up to any positive number. We would not, in practice, expect to find many theories with degrees of independence between zero and one: such "theories" are actually more complicated than that which they purport to explain. It is only when a theory acquires a degree of independence greater than one that it has those features we particularly value in a scientific theory. Moreover, it will be argued, the more above one it is, the more likely it is that it will lead us to correct predictions.

THE JUSTIFICATION OF THE PREFERABILITY OF PREDICTIONS MADE BY THEORIES THAT ARE HIGHLY INDEPENDENT OF THE DATA

In this section it will be argued that an *a priori* justification can be given for the claim that predictions made by theories that are highly independent of the data are to be preferred to those that are not. This justification is like the justification of induction given in the previous chapter, but contains aspects not present in the argument for induction.

It is very important to understand that in this section it will *not* be argued that independent theories are more likely to be *true*, in the sense in which a scientific realist might claim that a theory is true. A scientific realist, of course, is likely to claim that a (mature) scientific theory is true not merely in the sense of leading to correct empirical predictions, but in the sense of making claims that "correspond to the facts" about, for example, the behaviour and properties of theoretical micro-entities. But the argument to be presented in thwis section provides no justification for the claim that independent theories are more likely to be true in that sense. It is *only* argued that independent theories are to be preferred because they are more likely to make empirical predictions that subsequently turn out to be correct.

We have already noted that it is possible – or that it seems to be possible – to draw a distinction between the *novel* and the *familiar* predictive success of scientific theories. A theory enjoys novel predictive success if it successfully predicts instances of observable regularities different from those used to initially formulate it, whereas a theory enjoys merely familiar predictive success if it only predicts new instances of the same regularities as those on the basis of which it was initially formulated. As we proceed through the argument of this section, it might seem it only warrants us in preferring the *familiar* predictions made by a highly independent theory. But it will later be argued this is not so: the justification can be extended to novel predictive success.

The argument to be given relies crucially on the notion of the *contingent location of the data*. What follows immediately below is our *initial statement* of the argument for the preferability of highly independent theories (we will see later that it will need to be qualified if it is to be able to meet an apparently powerful objection).

1. The more independent a theory is from the data on which it is based, the less likely it is that the fact that the data exemplify this theory is a merely chance occurrence due to the contingent location of the data.
2. Therefore the more independent a theory is from the data, the more likely it is that the fact that it is exemplified by the data is *not* merely a chance occurrence due to the contingent location of the data.
3. Hence the more independent a theory is from the data on which it is based, the more likely it is that there is a propensity or law-like tendency for the data to exemplify that theory; that is, the more likely it is that the data conform to a law of nature corresponding to that theory.
4. Thus the more independent a theory is from the data, the more likely it is that if the data had been obtained from a different location, or under different conditions, they would still exemplify the same theory.
5. The more independent a theory is from the data, the more likely it is that any prediction it makes concerning the behaviour of the data in different locations will be successful.

The moves from step 3 to step 5 seem unproblematic. What needs to be defended is the correctness of the initial premise, step 1, and the moves from step 1 to step 3.

A DEFENCE OF THE INITIAL PREMISE OF THE ARGUMENT

We have already given a brief defence of the correctness of step 1 in our discussion of R1, R2 and R3. We noted that it was certain that any body of data should exemplify a rule as dependent as R1, that it was reasonably likely it should exemplify a rule as independent as R2 and that it was rather unlikely that it should exemplify a rule with the degree of independence of R3.

In defending the premise (step 1), we will initially appeal to our *intuitive notion* of a single explanatory component. A single explanatory component registers what we would intuitively regard as a single pattern embedded within some larger body of data. For example, we intuitively judge the clauses "Add the lowest even number twice" and "Add 8s and 4s alternately" to be *distinct* sub-rules, or distinct explanatory components. And it is this *intuitive* notion of distinct patterns, or distinct rules for generating data, that we will initially use in arguing for the correctness of the first premise.

Suppose we select three consecutive numbers at random. We find that they are all 2s. The *a priori* probability of three 2s in a row is rather low: assuming the numbers can only be single digits it is one in a thousand. The *a priori* probability of us selecting *any* three consecutive numbers the same all in a row is also pretty low: one in a hundred. But the probability of us selecting three consecutive numbers that obey some *reasonably simple pattern* is presumably rather high, although it we cannot attach a precise probability to it, because we do not know (or have not worked out) how many of the possible combinations would count as a "reasonably simple pattern". Also, of course, simplicity of patterns comes in degrees. Intuitively, "2, 2, 2" is simpler than "2, 4, 6", which is in turn simpler than "1, 4, 9". The author performed a search through a random selection of sequences consisting of three digits. It was found that about 20 per cent of them exemplified what we would regard as "very simple" patterns, while another 20 per cent exhibited "moderately simple" patterns. So there is a probability of about 0.4 that a sequence of three numbers should exemplify what we would ordinarily regard as at least a "moderately simple" pattern.

The situation is quite different if we turn our attention to longer sequences of numbers. Let us suppose that a sequence of six consecutive numbers were selected from some arbitrary location in larger sequence of data. What are the chances that a sequence of six such numbers would exemplify what we would regard as a single pattern, or be explicable by what we would intuitively regard as a single rule? Of a random selection of one hundred sequences containing six numbers chosen at random, none were found by the author to exemplify what we would regard as a single, moderately simple, pattern.

These simple observations enable us to develop a persuasive argument for the claim that the more independent a theory is from the data on which it is based, the less likely it is that the data should exemplify that theory by chance. Let us consider a hypothetical sequence consisting of six numbers. We have noted that the probability of the first three numbers in this sequence exemplifying what we would regard as a reasonably simple pattern seems to be about 0.4. The probability of the second three numbers also exemplifying some intuitively simple pattern or other (not necessarily the same as that exemplified by the first three) will also be about 0.4. Hence, the probability that the sequence of six numbers will be explainable by just two DECs is about 0.40 by 0.4 = 0.16. But now let us consider the probability that a sequence of six numbers should be explicable by *just one* DEC; that is by a theory with a degree of independence twice the first. If it were the same as the probability of it being explicable by a theory consisting of exactly two DECs, then we would expect a collection of one hundred 6-digit sequences to contain about 16 sequences that exemplified what we would intuitively regard as a single pattern. But, as we noted above, the author in fact found none. What this suggests is that 6-digit sequences that can be explained by a single DEC are much rarer than 6-digit sequences that can be explained by theories with two. And there is a quite straightforward reason why this should be so. For a 6-digit sequence to be explainable by a theory with two DECs, it is sufficient that the first three numbers in it exemplify what we would intuitively regard as one pattern, and for the last three numbers to exemplify a different pattern. But if the numbers are to be explained by a single DEC, the pattern exemplified by the first three numbers must the very same as that which is exemplified by the second three.

But now, if we appeal to our intuitive notion of a *single* pattern, we find that there are many – and, crucially, *more than two* – patterns that can be exemplified by three digits. For example, the sequences (4, 4, 4), (1, 2, 3), (9, 8, 7), (2, 4, 8) and (1, 4, 9) all exemplify what we would intuitively regard as single, but different, patterns. So if the first three digits in our sequence have exemplified pattern P, then there *is only one way* in which the sequence can be explicable by a theory with just one DEC, and that is if the second three numbers in the sequence also exemplify P. But there are *many ways* in which the second group of three numbers in the sequence can exemplify a pattern different from P. Therefore the chances of a 6-digit sequence being explicable by one DEC are lower than the chances of it being explicable by two, that is by a theory with twice the degree of dependence. Obviously, these remarks also apply to sequences of digits of any length.

In our discussion so far, we have focused on patterns in sequences of numbers. But the points made can easily be extended to data of other types. Suppose we have six Xs, and have observed the first three to all be green. This can be represented as the pattern "green, green, green". If we use our intuitive notion of "the same pattern", there is only one way in which the remaining three Xs can exemplify the same pattern, and that is if they are also "green, green, green". But there are *several* ways in which the second three could exemplify a pattern different from the first three. Some examples are "red, red, red", "blue, blue, blue" and so on. Or the sequences "green, blue, violet" and "green, yellow, orange" could be regarded as simple patterns: in the first case, the frequency associated with the light is increasing, in the second it is decreasing. No matter what the nature of the data, there are more possible ways in which the various parts of it can exemplify a larger number of patterns than there is for the whole of it to exemplify a single pattern.

In general, there will always be more ways in which a sequence can exemplify a theory with a lower degree of independence than one with a higher. Hence the *a priori* probability that any arbitrary body of data will exemplify a theory with a higher degree of independence will always be lower than the *a priori* probability that it will exemplify a theory with a lower degree of independence.

Now, let us suppose we have some very large body of data. The data might be a series of measurements the results of which are recorded as numbers. Let us suppose, for definiteness, that the body of data has a million individual items. We arbitrarily select some small subset of the data, and we find that the subset conforms to some simple pattern. For example, the pattern might be that each item of data is twice the one before it. Again, for definiteness, let us suppose we select six consecutive items of data about half way through the whole data set, and they conform to this pattern. The question arises: is it just due to chance that we happened to select a subset that conforms to this pattern; that is, is it merely due to the contingent location of the data that it happens to exemplify this pattern, or can we reasonably expect other samples drawn from the body of data to exemplify this same pattern? The argument given above shows that the more independent from the data the pattern is, the less likely it is that it should merely be due to the contingent location of the data that it should exemplify the pattern.

The above is, I think, about as close as it is possible to get to a rigorous justification of our initial premise (step 1 on page 68). It is not possible to obtain an entirely rigorous, analytic proof of the claim as long as we appeal to the intuitive idea of "the same pattern", which is a somewhat vague concept. Later in this chapter I attempt to make this idea clearer (see page 80ff.).

Let us now consider the move from step 1 to step 3. This is the move from:

> The more independent a theory is from the data on which it is based, the less likely it is that the fact the data exemplify this theory is merely a chance occurrence due to the contingent location of the data.

to the conclusion:

> The more independent a theory is from that data on which it is based, the more likely it is that there is a propensity or law-like tendency for data to conform to that law.

Now, we have already encountered, in the previous chapter, an inference that is rather similar to this. The previous inference was:

> This coin has been tossed a million times and it has come up heads every time. *Therefore* the coin is not fair; that is, there is a greater propensity for heads to come up than there is for tails to come up.

Intuitively, it seems that the inference to the greater propensity for heads is very reasonable. Is the inference involving the independence of theories similarly reasonable? It will be argued that it is, although a number of issues must be addressed before we are entitled to come to that conclusion.

OBJECTION: THE ARGUMENT FOR THEORIES THAT ARE HIGHLY INDEPENDENT OF THE DATA COULD ALSO BE APPLIED TO THEORIES EXPRESSED IN BIZARRE "GRUESOME" PREDICATES

In the remainder of this chapter we will consider objections to the above argument for highly independent theories. Although a variety of objections are considered, it will be argued that in all cases the replies to the objections are variations on the same underlying idea: ad hoc dependence is bad; independence is good. It will be argued that a notion of lack of ad hoc dependence applies not just to theories, but also to the predicates in which theories are expressed and to the "good-making" properties of theories. It will be argued that if we are consistent in preferring independence and eschewing dependence, the difficulties to be raised in the remainder of this chapter can be avoided.

Earlier (page 65ff.) it was argued that theories with a high degree of independence from the data also had an increased chance of subsequent empirical success. But neither that argument nor the discussion of components of data and theory in the previous section makes any mention of the types of predicates that can appear in the theories. Specifically, grue-type predicates are not forbidden. But, it may be objected, if theories expressed in contrived, "artificial" predicates are permitted, then it is easy to construct theories with a very high degree of independence from the data, but which we are sure would have no predictive success at all. We can illustrate this by considering again the sequence of numbers S1 given above. Let us say that a sequence has property Φ if and only if it has the number "2" in its first position, number "4" in its second position and so on, in conformity with S1. So if a sequence has Φ, it will have the number "46" in its tenth position. But let us define "Φ" so that a sequence is Φ only if it consists entirely of zeros for all positions after the tenth position. Clearly, if we have only seen the first ten members of S1, we can say "So far, S1 has had Φ." Moreover, the hypothesis that S1 has Φ is highly independent of the data, since it can explain all the data with a single component of theory. So, it may be objected, on the position defended here we are warranted in asserting that S1 will continue to have Φ, and therefore that all further members of the sequence will be zeros. Clearly, the same process could be applied to any body of data to construct a theory, of a high degree of independence, that explained that data, but which also made any bizarre empirical prediction we wished. But if this is so, then the notion of the independence of theory from data is plainly quite useless as a way of *explaining* the novel predictive success of actual theories.

Intuitively, what is wrong with the theory that all members of the sequence have Φ is that it is still highly ad hoc. It has all the ad hocness of rule R1 (page 61). But in this case the ad hocness resides in the *definition of the predicate* Φ. We are perhaps inclined to say that the high degree of *dependence on the data* is in the definition of Φ. And so we are led to the question: Under what conditions can the definition of a predicate be said to have a high or low degree of "independence from the data"?

INDEPENDENCE OF PREDICATES

The reader may recall that in Chapter 2 it was stated that, although the position defended in this book does not presuppose scientific realism, it does presuppose a form of metaphysical realism. One (rough and ready) definition of metaphysi cal realism is that the world has its existence and features *independently* of mind, language or other forms of representation. There is, that is, a notion of "inde pendence" that features in the concept of metaphysical realism. On the face of it, however, this (metaphysical) notion of independence might seem quite uncon nected to the notion of independence (in the sense of lack of ad hocness) with which we are here concerned. But in this section it will be argued that in fact the two notions of "independence" are connected. The notion of independence that appears in metaphysical realism, it will be argued, must be incorporated into our notion of independence of theory from data.

From the point of view of naïve common sense, the members of the class of green things are more similar to each other than are the members of class grue things. Of course, the thesis could also be defended that the members of class grue things *are* all similar to each other – they are all *grue*. But, we can surely ask, are the grue things *really* as similar to each other as the green things? This is, of course, a very well-worn area of philosophical inquiry, and I will not have anything new to say about it here. But one available position on this set of issues – developed in particular by David Armstrong (1978: vol. I) and David Lewis (1983: 347–77) – says that, for example, the green things are more similar because *there is a relation of mind- or language-independent similarity between its members*. This position is metaphysical realist in that it asserts that there are some things (relations of simi larity and difference) that exist independently of language and thought. But it is not scientific realist in that it does not assert the existence of the unobservable entities postulated by scientific theories.[2]

It should also be noted that the form of metaphysical realism presupposed here is rather weak. It is very much weaker than what is *usually* defined as metaphysical realism. Michael Dummett, for example, identifies realism with a commitment to the law of bivalence; but the existence of mind-independent relations of similar ity and difference need not commit us to realism in that sense.[3] Hilary Putnam has suggested that a metaphysical realist is committed to exactly one true theory of the world, but saying there are mind-independent relations of similarity and difference does not commit us to that.[4] It is much weaker than the doctrine that Michael Devitt calls "Realism" in the classification of positions developed in his *Realism and Truth* (1991: 302–3) and is also weaker than Armstrong's realism about universals.[5]

Let us introduce here the idea of a *basic natural predicate*. A term is a basic natural predicate only if it has its extension fixed by ostending some finite collection of samples, and then explicitly or implicitly asserting that the term refers to all objects that are of the same relevant kind as the ostended samples. It is a familiar thesis that typical natural kind terms are basic natural predicates in this sense, but we need not enter here into the debate over just how many natural kind terms fit this model. If the term is to be a basic natural predicate in our sense, the relation of being of the same kind as the ostended samples must also be a mind-independent or language-independent relation. Clearly, then, the elements of the extension of a basic natural predicate will bear a mind- or language-independent relation of similarity to each other.

The notion of a basic natural predicate enables us to introduce a notion of the degree of independence of a predicate. If a predicate has only a finite number of elements in its extension, then its degree of independence is the ratio of the number of elements in its extension to the number of components in the definition of the predicate, when that definition is stated in basic natural predicates. Of course, this does not allow us to assign different degrees of independence to two predicates both of which have infinite extensions. But we may say that if P and Q both have infinite extensions, then P is more independent than Q (i) if P and Q are co-extensive and P has fewer components in its definition when the definitions of both predicates are stated in basic natural predicates, and (ii) if the number of components in the definitions of P and Q are the same when stated in basic natural predicates, but the extension of Q is a proper subset of that of P.

There are two possible ways of eliminating theories expressed in terms of predicates such as "grue" or "Φ". One way is to stipulate that in order for a theory to have a high degree of independence from the data, not only must it have a high ratio of components of data to dependent explanatory components of theory, but the predicates used in the theory must also have a high degree of independence. Another possible way is to stipulate that the degree of independence of a theory from the data is to be assessed only after all the predicates in the theory have been defined in terms of basic natural predicates. But, either way, theories employing terms such as "Φ" or "grue" would be ruled unsatisfactory.

It is worth noting that unless we in this manner restrict the types of predicate that can appear in a theory, the first premise in the defence of independence theories (as given on pages 65–7) is not true. This first premise, we recall, said that the higher the degree of independence of a theory, the lower the *a priori* probability that any actually obtained body of data should, merely by chance, conform to a theory of that degree of independence. But if we allow our theories to be stated in grue-type predicates, this premise is simply false. We can always formulate a "theory", using such predicates, that can "explain" any body of data of any degree of complexity with a *single* component of theory. The probability of there being such a "theory" is 1. Only if we rule out grue-type predicates do we have an assurance that the regularities we have discerned exist "out there" independently of us. Only if we restrict ourselves to predicates that refer to mind- and language-independent relations of similarity does it become *hard* to discern simple patterns in the data, and only then can we be justified in saying that the presence of the simple pattern could not be due to chance.

This leads to a conclusion that might seem paradoxical, or at least surprising. We cannot "directly see" how nature is; we must use criteria such as simplicity to help us go beyond these parts and obtain knowledge of those less accessible parts. On the view to be developed in this book, it is not simplicity *per se* that plays this role, but the independence of theory from data. But if there are to be theories that are independent of the data, there must be mind- and language-independent features of reality, such as mind- and language-independent relations of similarity and difference. If we are permitted to construct theories using any predicates whatsoever, including the most ad hoc, grue-type predicates, it is easy to get a theory that can explain the data, and easy to construct a theory that makes any prediction we please. But if we restrict ourselves to predicates that have as their extensions natural classes of objects, then it becomes hard to construct an independent theory that can account for the data. And in such cases we are entitled to say it could not be due to chance that the data conforms to the theory. It is the use of mind-independent predicates that makes it possible for our theories to get "traction" on the observable world, and latch on to regularities in the data that hold independently of language and concepts. Hence, on the view to be defended here, it is mind- and language-independent features of reality that actually make it possible for us have knowledge of aspects of the world that goes beyond what we have observed. Putting it perhaps a little over-simply, the paradox is that it is only because there are mind-independent features of reality that the mind can get a grip on the less accessible features of reality. We can only get a grip on something larger than our hand if that which we are grasping to some extent resists our grasp.

OBJECTION: THE VIEW DEFENDED HERE ASSUMES THAT "NATURAL" PREDICATES ARE BASIC

It might be objected against the view advocated here that it implicitly assumes a particular vocabulary as basic or fundamental. If the terms we regard as natural – terms such as "green" and "blue" – are taken as fundamental, then perhaps the definition of independence of theory from data used here will result in theories we intuitively regard as "good" being selected. But if it is rather terms like "grue" and "bleen" that are taken as fundamental, the result will be very different. After all, as Nelson Goodman has famously argued, the terms "green" and "blue" are only basic or fundamental relative to a particular choice of vocabulary. If we adopt a vocabulary in which "grue" and "bleen" are fundamental, then it is the terms that we regard as natural that turn out to require (complex) definitions. For example, if "grue" and "bleen" are taken as fundamental, then the definition of "green" may be "grue before D-Day, or bleen thereafter". And so, it may be claimed, our definition of independence will only issue in a satisfactory way of discriminating between good theories and bad if our basic vocabulary is chosen appropriately.

However, at least on Lewis's theory of reference, this is not so. On Lewis's theory, it is simply not possible to choose a vocabulary in which "grue" and "bleen" are basic. This can be brought out by considering more closely just how his theory works. Suppose a speaker is using a term "g". In an initial attempt to specify the extension of "g", the speaker stipulates it is to be a colour term, and points to a number of green objects, saying they are all g. Assume these to be the only facts about the speaker (or the community of which they are a member) constraining the extension of "g". To what class of entities will "g" refer? On Lewis's view, it will be to the most natural class consistent with the other constraints (that "g" is a colour term, and has so far only been applied to green objects.). And, presumably, this means the extension of "g" will be the set of green things. But, on Lewis's view, "g" refers to the green things not because we have given it a definition ensuring it so refers. In the absence of a definition, or anything else further constraining its referential properties, "g" will "automatically" refer to the class of green things.

Of course, we could make "g" refer, for example, to the class of grue things. But, a definition would be required to get "g" to do that. And that definition may just be the standard definition of "grue". But in such a case, "g" would not be a definitionally fundamental or basic term. On Lewis's view, if "g" is fundamentally definitional or basic, it will "automatically" take as its extension the most natural class available, and that will be (presumably) the class of green things.

What this shows is that, on Lewis's theory of reference, a language in which the terms "grue" and "bleen" are fundamental is not a possibility. Definitionally fundamental terms, on Lewis's view, automatically take natural classes of things as their extensions. Speakers can get terms to refer, for example, to the set of grue things or bleen things, but only by making those referring terms non-fundamental.

AN EPISTEMIC PROBLEM: OUR KNOWLEDGE OF MIND-INDEPENDENT RELATIONS OF SIMILARITY

An *epistemic* problem arises concerning our knowledge of the extensions of the predicates used in our theories. According to the view advocated here, our predicates take as their extensions classes of objects that bear mind-independent relations of similarity to each other. These relations of similarity hold independently of human perception, knowledge, verification or linguistic representation. So to say that two objects A and B are similar in this sense does not entail that the assertion "A is similar to B" is currently believed to be true, or even that it would necessarily be believed to be true under "epistemically ideal conditions". But then a problem arises: how can we *know* whether an object does or does not fall under the extension of one of our predicates? And, more specifically, how can we recognize whether or not an event is or is not of the sort that has been predicted by our theory?

We can bring out the problem with a specific example. Suppose, for example, our theory predicts that the next leaf we see will be green. Putting this in the "formal mode", we may say our theory predicts the next leaf we see will be a member of the extension of the predicate "green". What must we do to recognize whether this prediction has been confirmed or refuted? To consider a particular example, what must we do to tell if a given observable object O confirms or refutes our prediction that it will be green? If "green" takes as its extension a class of things that bear a mind independent relation of similarity to each other, we must recognize that the object O bears a particular mind-independent relation of similarity to some *other* objects. We must at least, it seems, recognize that O bears this mind-independent relation of similarity to the objects we typically take as paradigm cases of green objects, or to those objects that were used to initially fix the reference of "green". But then the question naturally arises: "How do we manage to do this?" If the relation of similarity is mind-independent – if it holds independently of our best beliefs about similarity – how do we know O is relevantly similar to the other objects?

It will be argued that a possible solution to this problem can be obtained by naturally extending some aspects of Lewis's theory of reference. Lewis, we recall, holds that our general terms take as their extensions the most natural, or objectively similar, classes of objects compatible with the other constraints on reference. It is suggested that we can solve the epistemic problem with which we are here concerned if we extend Lewis's idea to our perception of relations of similarity between objects.

Let us suppose that the reference of "green" was initially fixed, or "grounded", by ostending the members of a set S of green objects. Let O_1 be a green object not in set S and let O_2 be a red object. It seems clear that we can *see*, or *perceive*, that O_1 is relevantly similar to the members of S, while O_2 is not similar. We can see that they share "greenness". But, of course, it might also be maintained that there is a way in which O_2 *is* similar to the members of S. We could appropriately define a term "gred" (by analogy with "grue") and say that the elements of S and O_2 are similar because they are gred. Nonetheless, it is, at least with respect to colour, clear we do not perceive O_2 to be similar to the members of S.

It is suggested that, following Lewis's lead with reference, a theory of similarity-perception can be developed that has as one of its consequences that we perceive that O_1, but not O_2, is similar to the members of S. The intuitive idea is quite simple but needs to be stated clearly.

First some concepts need to be introduced. Can O_2 and the members of S be said to be, in some sense, similar since they share "gredness"? Does a "similarity relation" of *some* sort hold between them? We will say that a *broad* similarity relation holds between O_2 and the members of S. In general, a broad similarity relation holds between two objects X_1 and X_2 if it possible to construct a predicate P (where this may include grue-type predicates) such that P applies to both X_1 and X_2. Of course, broad similarity relations in this sense will always hold between any two arbitrarily chosen objects. To say that a broad similarity relation holds between X_1 and X_2 does not imply that we would be intuitively inclined to judge them as similar, and neither does it imply that they are objectively similar.

The notion of "broad" similarity is meant in an inclusive sense: X_1 are "broadly" similar if "artificial" or "non-natural" predicates apply to both of them: for example, if they are both grue or both gred. But they are also broadly similar if *objective* relations of similarity hold between them. They are broadly similar if they are both green, or red, or golden, metallic and so on.

There is another preliminary issue that needs to be considered. We generally judge objects to be similar or dissimilar in certain *respects*. A red circle and a red square are similar with respect to colour but not with respect to shape. If a speaker were asked to judge whether two such objects were similar, they might reply, "Similar in what way?" We will, for the moment, speak of similarity and dissimilarity relations of a given *category*: similarity with respect to colour is a similarity relation of one category; similarity with respect to shape, another. Our *initial* statement of our theory of similarity-perception will make use of the notion of categories of similarity, but it will later be argued it is possible to dispense with this notion.

Suppose we are confronted with two red objects A and B. There will be very many broad similarity relations between them; there will also be many broad relations of colour-similarity. Both of them, for example, will be red. But by inventing the appropriate predicates, we could specify many other broad relations of colour-similarity that hold between them. For example, they will both have the colour property of being neither orange nor pink. But we can say that the respect in which we perceive them as being similar is that of both being red. And the class of red things is a more objective or natural class than the class of things that are neither orange nor pink. This suggests the following (initial) account of similarity-perception:

> The similarity relation of a given category that is the *object of our perception* in a given observable situation is, from among the set of all similarity relations of that category present in the situation, the one that is most objective or natural.

It is worth stressing the relation of this account to Lewis's theory of reference. Lewis says that the extension of a given referring term K is the most natural class C compatible with the other reference-determining factors. For Lewis, this does not mean there is some additional fact about our psychological make-up that brings it about that we refer to C rather than some other class. It is rather, for Lewis, a fact about *the nature of reference itself* that this is the class to which we refer. Similarly, it is not being claimed there is some fact about our psychological make-up, or our perceptual apparatus, that brings it about that we perceive the most natural similarity relation. Rather, it is suggested that it is a fact about the nature of the intentional relation that holds between our act of perception and that which we perceive that the *most natural* similarity relation, compatible with the other factors, is the object of our perception.

There is one clause in Lewis's account of reference that requires our attention. For Lewis, a term T takes as its referent the most natural class *compatible with the other reference-determining constraints on T*. These "other determining constraints" might include the speaker's psychological state, dispositions to linguistic behaviour and causal links to various parts of the world. And it is surely permissible to appeal to the notion of "other reference- or object-determining constraints" in the case with which we are presently concerned. In a given perceptual situation, the similarity relation that is the object of a person's perception will be the most natural similarity relation compatible with the other constraints that exist , in that situation, on what they might be perceiving. What might these "other constraints" be? A natural answer might include the *type* of similarity that the perceiver is, in that situation, disposed to focus on. Presented with a number of coloured shapes, for example, a speaker might be disposed to focus on the ways in which their colours are similar, or they might be disposed to focus on the ways in which their shapes are similar. Which *respect* in which the objects are similar will be the focus of the speaker's attention is, plausibly, determined by the speaker's psychological state. And the perceiver's psychological state can be included in the "other constraints" on that which can be the object of the person's perception. This enables us to state the account of similarity perception we will use here:

> The similarity relation that is, in a given situation, the object of perception of a perceiver P is, from among the broad similarity relations present in that situation, compatible with the other constraints on similarity perception, the most natural of the similarity relations.

Let us see how this account applies to the problem with which we are here concerned. Suppose a theory predicts, for example, that in a particular experiment some liquid will turn green. A scientist performs the experiment, and observes the colour of the liquid. On the account advocated here, the claim that the liquid turns *green* is true just in case an objective, mind-independent relation of similarity holds between the objects taken by the speaker as paradigms of green things (or, alternatively, the things used to fix the reference of "green") and the sample of liquid the scientist perceives in the experiment. But how is the scientist to be able to recognize that the liquid now before him, and the paradigm cases of "green", are similar in *this particular way* if this relation of similarity is mind-independent? The answer, on the view developed here, is not that there is something about the scientist's act of perception, or psychological state, that is sufficient to pick out this relation. Rather (adapting from Lewis) it is a part of *the nature of the intentional relation* between the perceiver and the world that the similarity relation that is the object of the scientist's perception is the most natural one compatible with the other constraints present. And so, if greenness is the most natural relation between the paradigm cases of green and liquid now before the scientist, this is the relation of similarity which, on the account offered here, will be the object of the scientist's perception.

The problem with which we have been concerned in this section is "How can scientists recognize whether a theory has been confirmed (or refuted) if that recognition requires an ability to tell if an objective, mind-independent relation of similarity holds between one observable situation and others?" The solution offered here has been to develop a theory of similarity-perception. The theory is a natural extension of David Lewis's theory of reference. On this theory, it is a feature of the nature of the intentional relation between our act of perception and the world that we perceive the most natural similarity relation.

OBJECTION: NOT ALL PREDICATES OF SCIENCE ARE NATURAL PREDICATES

It is a consequence of the view advocated here that the terms of science take natural, or objectively similar, classes of things as their extensions. But, it may be protested, not all predicates of science are like that. Some terms used in science are not like "natural" predicates, such as "green", but seem to be rather more like terms such as "grue".

Modern physics seems to be a source of terms that, at least to common sense, seem to be more like "grue" than "green". For example, according to current understanding, neither mass nor energy is conserved, but the "hybrid" quantity "mass-energy" *is* conserved. And, on the face of it at least, this "hybrid" term seems rather more like "grue" than "green". Or, to take another example, modern physics tells us that a photon is neither entirely like a particle nor entirely like a wave: it is said to be a "wavicle". Again, science seems to be employing a term rather more like "grue" than "green". So, it might be protested, science does not (at least exclusively) consist of the "natural" terms claimed here.

However, it will be argued that this objection is in error: appearances to the contrary, the terms "mass-energy" and "wavicle" may in fact be entirely "natural" terms. The crucial point to note in this context is that if a term is a *theoretical* term, or has a theoretical component to its meaning, the class of things to which it refers might, *at the observational* level, look rather like a non-natural class. Consider, for example, the term "carbon". This term includes in its extension such observationally different things as lumps of coal and diamonds. But this does not mean that "carbon" is a term like "grue". The class of things to which "carbon" refers *is* a natural class: it is the class of things consisting (more or less entirely) of atoms with atomic number six. But the relevant similarity relation between these things exists at the theoretical level: observationally they may be quite diverse.

This suggests at least a possible strategy for dealing with terms like "mass-energy" and "wavicle". Let us consider "mass-energy" first. Special relativity tells us these are interchangeable quantities. It seems to be at least a permissible interpretation of this that mass and energy are two different manifestations of the same underlying quantity, which is termed "mass-energy". On such an interpretation, the term "mass-energy" does not take a non-natural class of things (quantities of mass *and* quantities of energy) as its extension; rather, it takes a *natural* class as its extension. The property shared by all members of this class is a *theoretical* property, which we refer to by means of the "hybrid" *term* "mass-energy". But although the *term* is "hybrid", the theoretical *property* shared by all members of this class is the same.

The interpretation of "wavicle" is similar, but slightly different. Photons, for example, seem under some circumstances to exhibit wave like properties and under other circumstances to exhibit particle-like properties. A *given* photon might, under some circumstances, behave like a wave, but under other circumstances, it will behave like a particle. But it is the *very same* particle that would behave in these different ways. And this is also true of each and every photon. *Every* photon behaves in some circumstances like a particle, and others like a wave. We can say that every photon has *the same* dispositions to behaviour; it is just that each one of them behaves in different ways under different circumstances. And so we can say that the class of photons is a natural class: every member behaves in the same way, it is just that their behaviour differs, in a very surprising way, from one experimental situation to another.

As with "mass-energy", the term "wavicle" is a hybrid *term*. But it takes as its extension what is in fact a natural class of objects.

In summary, the terms "mass-energy" and "wavicle" do not constitute counter-examples to our claim that the terms of science take as their extensions *natural* classes. Like "carbon", they might refer to things that, observationally, display a variety of appearances, but which, at the theoretical level, in fact bear a relation of objective similarity to each other.

OBJECTION: RULING OUT THEORIES EXPRESSED IN GRUE-TYPE PREDICATES FAILS TO RULE OUT ALL COUNTER-INTUITIVE CASES

On the view advocated here, the reason why highly independent theories are more likely to enjoy predictive success is the fact that it is *highly improbable* that a body of data should by chance conform to a highly independent theory. But, of course, independence is not the only improbable property of explanatory theories: it is easy to specify other highly improbable properties. If the defence of highly independent theories given here were sound, then we ought to expect theories with *any* highly improbable property to have an increased chance of predictive success. It is natural to object, however, that there are some highly improbable properties of theories that we most definitely would not expect to be associated with an increased chance of predictive success. Let us suppose, for example, we found that a theory that explained some body of data also happened to be isomorphic with a newspaper article on sheep farming in New Zealand. Let us also suppose that it is *a priori* highly improbable that the theory should be isomorphic with the article. Then, it may be objected, it is a consequence of the view advocated here that we could argue that there exists a tendency for the data to conform to the theory with this property, and hence that there is a good chance that it will continue to do so. That is, on the view advocated here, not only do we have a reason to use theories that are *simple* or *independent* to guide our predictions: we also have the same reason to use theories that are, for example, isomorphic with articles on New Zealand sheep farming. And this, it may be objected, it surely a *reductio ad absurdum* of the view.

So we are confronted with this problem: "If the argument purporting to justify a preference for theories with independence from the data works, why can't an argument of exactly the same form be used to justify a preference for theories with the highly improbable property of being isomorphic to an article on New Zealand sheep farming?" Intuitively, perhaps our initial response to this case is that it resembles the case of "grue" and other ad hoc constructions. But in this case the ad hocness does not lie in the definition of the predicate(s) that appear in the theory, but rather in the specification of the highly improbable property the explaining theory is said to possess.

This can perhaps be brought out more clearly if we consider the difference between the predicates "is independent of the data" and "is isomorphic with an article on New Zealand sheep farming". In the former case, we have a prior, independent reason for the preferability of such theories: the argument given in pages 67–8. But in the New Zealand sheep farming case, this is not so: we had no prior reason for preferring theories with this specific property. The predicate is merely constructed in an ad hoc manner in response to the observed properties of the theory itself. Moreover, the specification of the structural property of being isomorphic with the particular article on New Zealand sheep farming, if spelt out in basic natural predicates, would qualify as a highly dependent predicate.

This suggests a natural way of saying why the inference concerning New Zealand sheep farming is not warranted: the highly improbable property is constructed ad hoc from the characteristics of the theory itself. It is suggested that, in general, the inference to the likely future success of a theory is warranted only if the highly improbable property of the theory is not itself highly dependent on the observed features of the explaining theory.

Let us test this claim against an example. Suppose we have a deck of cards. We do not know what has brought the deck of cards to be in the specific order it is: perhaps it has been deliberately arranged or perhaps it has been shuffled in to a "random" order. We then examine the order of the cards. Two possible orders are as follows:

1. The cards are in the order two of diamonds, three of diamonds, four of diamonds, ... , ace of diamonds, two of hearts and so on, through to the ace of spades; that is, they are in an ascending order of "value".
2. Some purely "random" order (perhaps the first four cards might be seven of clubs, two of diamonds, jack of clubs, four of spades).

In the second case, we could construct a predicate Ψ where, by definition, a pack has Ψ iff the first card is a seven of clubs, the second a two of diamonds, the third a jack of clubs and so on for all the cards in the second deck. Obviously, it is highly unlikely that any deck chosen at random would have Ψ. But, equally obviously, we would not be justified making the inference that since Ψ is such a highly improbable, it could not be due to chance that this pack should have Ψ. And neither would it be a justified inference on the view advocated here. The definition of the predicate Ψ is constructed ad hoc in response to the actual order of the cards. It would clearly qualify as a highly dependent predicate on the definition given above: it would contain a separate clause for each card in the stack and their position. So, on the criterion suggested here, the inference to the conclusion that it is not due to chance that the pack has Ψ is blocked.

OBJECTION: THE REPLY GIVEN IN THE PREVIOUS SECTION DOES NOT RULE OUT CASES IN WHICH THE "BIZARRE" IMPROBABLE PROPERTY OF THE EXPLAINING THEORY IS NOT POSTULATED AD HOC

It may be objected that in the previous section we failed to distinguish between two importantly different cases:

(i) The case in which the very improbable property of the explaining theory is only postulated in an ad hoc manner in response to the observed features of the explaining theory; and
(ii) The case in which the very improbable property of the explaining theory is not postulated ad hoc in response to the observed features of the theory, but specified in advance.

If a rare property of the theory (such as being isomorphic with the New Zealand sheep farming article) had been discovered after the theory had been formulated, and on the basis of a study of the structure of the theory, then the argument for the preferability of theories with that property fails to go through. But if the rare property had been specified in advance of the theory being formulated, or independently of it, then matters are not so straightforward. Recall that there are two dimensions to dependence or ad hocness: the post hoc dimension and the complexity dimension. If some predicate of an explaining theory has been specified in advance, then, even if it is highly complex, it would lack the post hoc dimension of dependence, and so fail to qualify as a highly dependent predicate on the view given here. And so, if the predicate were specified in advance, it appears that the inference to the conclusion that there exists a tendency for the data to conform to a theory with this property would not be blocked.

My response to this objection is to "grasp the nettle" and argue that, if the improbable property was specified independently of a knowledge of the structure of the explaining theory, it is not unreasonable to assert that a theory with that property will have an increased chance of predictive success. The plausibility of this response can, I think, be brought out by considering the following hypothetical example. Let us suppose that, high in the Himalayan mountains, an unusual form of water has been discovered, the boiling point of which seems to vary from time to time. Also suppose that, as far as scientists have been able to discover, the boiling point of the liquid does not depend on factors such as pressure or the purity of the sample: the variation has no apparent cause. Now let us assume that an eccentric Hindu scientist decides to study the water. This scientist has, prior to and independently of his decision to study the water, a belief that the *Bhagavad Gita* holds the key to understanding the laws of nature. He proceeds to heat the water up on a succession of different occasions and record the temperature at which it boils. He then translates this record of temperatures into letters according to the following rule: 100°C = "A", 101°C = "B", 102°C = "C", and so on. When his data are converted into letters in this way, it reads, "On the field of Truth, on the battle-field of life", the opening words of the *Bhagavad Gita*. Assume the scientist carries out, let us say, a thousand observations, and establishes that this pattern holds good for the first thousand letters of the *Bhagavad Gita*. He then advances the hypothesis that all observations, including future observations, of the water will conform in this way to the *Bhagavad Gita*.

It is a consequence of the view defended here that the scientist's hypothesis is a good one. It is clearly extremely unlikely that it should merely be due to chance that the data obtained should be explainable by a theory isomorphic with the *Bhagavad Gita*. Moreover, it is wrong to protest that the property of being isomorphic to the *Bhagavad Gita* is highly dependent on the data, since in this case we have assumed the scientist had specified the *Bhagavad Gita* prior to, and independently of, making his observations. The specified property does not count as dependent in our sense. So, on the view defended here, the inferences remain "unblocked" and so we ought to expect the scientist's hypothesis to have a good chance of making successful predictions in the future. But is this such a counterintuitive result? Of course, under these conditions we would say something very queer must be going on, but if a very large number of observations had been made in conformity with this pattern, it does not, I think, seem implausible to assert that there would be an increased chance that future observations will conform to this pattern. So here it is suggested we "grasp the nettle" and accept that under these conditions a theory would have an increased chance of future predictive success.

EPISTEMIC AND PROPENSITY NOTIONS OF PROBABILITY

It is customary to distinguish between epistemic and objective or physical notions of probability. Consider, for example, a coin. If we know nothing about the distribution of mass in the coin or about its past behaviour when tossed, then there is a sense in which we may have no reason to prefer either the hypothesis that the coin will come up heads or that it will come up tails. The hypothesis that it will come up heads is as credible as the hypothesis it will come up tails. So, we may say that heads and tails are equally probable in the "epistemic" sense of probable. If, however (unknown to us) the coin is heavier on one side than the other, then there may be a greater propensity for one side to come up rather than the other.

Epistemic probabilities may be known relatively *a priori*: we may know that a heads and a tails are equally epistemically probable merely if we are ignorant of the distribution of the mass in the coin. However, epistemic probabilities cannot be used to explain observed frequencies. If the coin is tossed one hundred times and it comes up heads about fifty times, we cannot explain why it in fact came up heads about fifty times by saying that, in our ignorance, the two outcomes were, as far as we knew, equiprobable. Propensities, on the other had, can play an explanatory role. We can explain why the coin came up heads about fifty times by saying, for example, that due to a symmetrical distribution of mass in the coin, the two sides have an equal propensity to come up. But, of course, the existence of a propensity in a coin is not something we can know *a priori*. In summary, epistemic probabilities may be knowable *a priori*, but they cannot be used to explain observed frequencies, while propensities can play an explanatory role, but cannot be known *a priori*.

It might be thought that these points create a difficulty for the point of view defended here. Our aim is to provide an explanation of certain cases of surprising empirical success in science. It has been argued that having a high degree of independence from data gives a theory an increased probability of having correct empirical consequences. So, here the notion of probability is playing an explanatory role. But it has also been argued that is *a priori* that if a theory has a high degree of independence from the data, then it has an increased chance of empirical success. And so it might be objected that perhaps here *a priori* assessments of probability are illegitimately also playing an explanatory role.

However, on closer inspection, this objection fails. In broad outline, the thesis defended here is that if some body of data is explainable by a theory with a high degree of independence from that data, then it is epistemically likely that there exists a tendency or propensity for the data to continue to conform to that theory. If it turns out that the data does continue to conform to the theory, then this is to be explained by saying that the data has a propensity to conform to the theory. So both epistemic and propensity notions of probability are used in the overall explanation.

The continued empirical success of the theory is to be explained by saying the data has a propensity to conform to it. So it is a *propensity* notion of probability that plays the explanatory role, as it should be. But note that the existence of this propensity cannot be established *a priori*. We can only establish the existence of this propensity by establishing that, as a matter of fact, there exists a body of data that conforms to a particular highly independent theory. This is something that can only be established empirically, or *a posteriori*. So no claim is made here that the existence of the explanatory propensity can be established *a priori*.

The relevant claim that is *a priori* is the following conditional:

> It is epistemically likely that *if* a body of data conforms to a highly independent theory, *then* there exists a propensity for the data to conform to the theory.

This conditional, however, is by itself not sufficient to explain the success of a theory: it needs to be empirically established that the antecedent clause of the conditional is true.

It ought to be noted that the conditional sentence claimed to be *a priori* is, plausibly, neither analytic nor necessary. But it was argued in Chapter 3 that there can be (defeasible) *a priori* claims that are neither analytic nor necessary.

THE INDIVIDUATION OF COMPONENTS OF THEORY AND COMPONENTS OF DATA

One theory is more independent of the data than another if it has a lower ratio of dependent explanatory components of theory to components of data than the other. Plainly, then, if we are to be able to say that one theory is more independent of the data than another, we will at least need to be able to say when one theory has *more* components of theory than another, and may even need to be able to *count* components of theory and data. And certainly, if we are to assign actual numbers to the degrees of independence of a theory, we will need to be able to count components of theory and components of data. But if we are to *count* components of theory and data, we need to be able to individuate them. We address ourselves to the problem of individuating components of theory and data in this section.

We will begin by considering the individuation of components of theory. There are, broadly speaking, two types of thing that can play an explanatory role. These are *laws* and *statements of "initial conditions"*. Our primary concerns in this section will be with comparing the degrees of independence of different sets of laws and with individuating the components of theory that exist within systems of laws. The problem of individuating components of theory in "statements of initial conditions" will also be briefly addressed.

In considering the problem of the individuation of components of theory, there are two things that are worth bearing in bearing in mind:

The notion of the independence of theory from data must be a   (i)
measure of the extent to which *it could not be due to chance* that the
data conforms to that theory; and

In at least the cases we have considered so far, a single component   (ii)
of theory corresponds to what we would intuitively regard as a
single pattern.

Clearly (i) will serve as a constraint on any satisfactory way of individuating components of theory: it *must* be unlikely for highly independent theories to have been
exemplified by chance if our argument for the preferability of independent theories
is to go through.

Let us begin by considering our intuitive notion of a pattern. A pattern is always
a *correlation* of some sort: I venture out in to my backyard and the dog next door
always barks; whenever sugar is placed in water, it dissolves; whenever electricity is
passed through a wire, the wire is heated. It is a familiar idea that such correlations
can be represented by unrestricted universal generalizations of the form "for all
x, if P(x) then Q(x)". So we may, as a first approximation, say that an unrestricted
universal generalization counts as a single component of theory.

However, as it stands, this won't quite do. It is easy to construct unrestricted
universal generalizations that certainly do not possess the characteristics we want
from a single component of theory. This can be brought out by considering the
following example. Suppose we observe that certain substances emit "Π-waves",
and we wish to discover which substances do this. We note that carbon, boron,
water, eucalyptus wood, plastic and aluminium all emit Π-waves. We then define
a predicate "pigenic" as follows:

x is pigenic if and only if x is carbon, or x is boron, or x is water, or   (1)
x is eucalyptus wood, or x is plastic, or x is aluminium.

This enables us to state the following generalization:

For all x, if x is pigenic, then x emits Π-waves.   (2)

However, we would be reluctant to count (2) as a single component of theory: it has
not got on to what we would regard as a single pattern. It rather describes what we
would intuitively regard as six distinct patterns or regularities which are given the
superficial appearance of a single regularity by the definition of the term "pigenic".

A natural way of accommodating this difficulty is to stipulate that if an unrestricted universal generalization of the form "For all x, if P(x) then Q(x)" is to
count as single component of theory, then the terms "P" and "Q" must register
what we would regard as objective, language-independent similarities. However, it
might be feared that stipulating that predicates must register objective or language-
independent similarities would violate the *accessibility requirement*: that it be easier
to tell if a theory is independent of the data than that it will turn out to enjoy the

forms of success exemplified in the phenomena. To say that a similarity is *objective* is to say that it holds independently of our choice of predicates or other means of representation. But then, it may seem to follow, to say that a similarity is objective is to say that it is not directly accessible, and, indeed it is not entirely clear how we would go about establishing that a relation of objective similarity held.

One natural way around this difficulty is to appeal to an account of reference explicitly developed by David Lewis (1983: 374–6). Lewis postulates that it is an analytic fact about reference that our general terms take as their referents those classes, compatible with all the other constraints on reference, that have the highest degree of objective similarity between their members. The important point is that, in order to explain why our terms refer to classes of objects the members of which are objectively similar, we do not need to postulate some special power of the human mind. In particular, we do not need to say that the human mind has some special power to reach out and identify the objective or language-independent similarities. Rather it is, to use a phrase due to Putnam (1981: 25), the objective "world itself that completes the job of fixing the extension of our terms".

We can here again appeal to the notion of a "basic natural" predicate. As noted earlier, the members of the extension of a basic natural predicate will be objectively similar. Moreover, it is possible for us to specify what is to count as a basic natural predicate in a way that is compatible with the accessibility requirement. A basic natural predicate has the feature that it is easier to *learn its meaning by ostension* than it is to learn the meaning of a "non-natural" term. In order to teach by ostension the use of a term such as "green" or "aluminium", it is necessary to show a speaker some samples of green things, or of some aluminium, and perhaps also explain to them the general category of thing to which it belongs ("green is a colour; aluminium is a metal"). But in order to teach a non-natural term by ostension, it is necessary to go through this process *several times*. To teach the use of grue, it is necessary to teach the use of green, of grue and the date of D-Day. To teach the use of "pigenic" it is necessary to teach the use of "carbon", "boron", "water" and so on. So, it seems that straight terms can be identified by means of the ease with which their meaning can be taught by ostension.

In summary, it is suggested that a single DEC must be couched in basic natural predicates, and that this can be done in a way that does not violate the accessibility requirement.

We have tentatively identified a single component of theory with a single, unrestricted, universal generalization. However, it is not too difficult to see that some difficult cases arise. Consider, for example, the formula for the strength of gravity around a point mass:

$$F_g = \frac{G.m}{r^2}$$

This can be seen as telling us about *two* patterns. First, it entails that if we select an arbitrary point mass P, the strength of the gravitational field around P will conform to a particular pattern; specifically, its strength will, at any location around P, always be proportional to the mass of P divided by the square of the distance of that location from P. We will call this the "first pattern" the law tells us about. But there is also another pattern: the law tells us that this first pattern will be exemplified, not just by the locations around the particular point-mass P, or about some subclass of all the point-masses: it tells us that this pattern will be exemplified by the locations around *all* point-masses. It is a generalization "twice over": first about the strength of gravity surrounding some particular point-mass, and second about *all* point masses. It is a "multiply quantified" statement.

Given that the law of gravitation asserts the existence of *two* patterns, ought we to count it as *two* components of theory, or ought we only count it as one? Since it asserts the existence of two patterns, and since we have suggested that a single component of theory ought to correspond to a single pattern, it might be thought that it should count as two components. But it will be argued that this is a mistake. First, let us note that although the law of gravitation tells us of two patterns, those patterns are related in a very different way to those described by R2 (page 61). R2 discerns two patterns in one body of data with, roughly, the first half of the data exemplifying one pattern and the second half exemplifying the other. Each pattern "covers", roughly speaking, one half of a particular body of data. But the two patterns asserted to exist by the law of gravitation are not related in that way. The first pattern covers *all* the locations around some arbitrary point mass; the second asserts that same pattern can be found around *all* point masses. The second pattern is a "meta-level" pattern, or a pattern in some class of lower-level patterns. It discerns a *single* pattern exhibited by all the locations around all the point masses. It is an unrestricted, albeit multiply quantified, universal generalization. It should, therefore, count as a single component of theory. And this is, of course, the result we intuitively want. While we feel that R2 is a rather bad "theory", the law of gravitation should surely count as a very good theory. By saying that it counts as just a single component of theory, we ensure that it does count as a good theory. So, I conclude, our initial suggestion that an unrestricted universal generalization is to count as a single component of theory remains viable.

Let us consider again the rule R2. Intuitively, this discerns two patterns in the data. (The first pattern was "add the lowest even number twice, then add the second lowest even number twice"; the second pattern was "add 8s and 4s alternately".) Let us refer to these two patterns as "$R2_1$" and "$R2_2$". The first of these rules held for the first five members of the sequence S1, the second for the second five. We can, therefore, represent these patterns as *restricted* universal generalizations. For example, $R2_2$ can be expressed as "For all x, if x is one of the second five members of S1, then it is given by the rule 'add 8s and 4s alternately'." This restricted universal generalization corresponds to what we would intuitively regard as a single pattern, so let us assert that a *restricted* universal generalization also corresponds to a single component of data.

So, in summary, a universal generalization expressed in straight predicates, whether restricted or unrestricted, corresponds to a single component of theory.

Let us now turn to the question of the individuation of components of data. At first it might be thought that a single component of data can be identified with a single observation. By a single observation I mean particular, dated event such as "At 4.02 pm the needle on the meter was observed to be pointing towards 5.2 on the scale." However, if our aim is to develop a measure of independence that is useful for comparing the degrees of goodness of different theories, this way of individuating components of data is inappropriate. When we are comparing the goodness of different theories, or theoretical systems, we do not count up the number of *single observations* that each theory can explain. We rather compare the number of *types* of observations, or *types* of tests, that can be accounted for or passed by the theories. And part, at least, of the reason for this is that we are generally confident that it is easy to make just about any theory pass an indefinitely large number of individual tests. For example, suppose we have observed that, under certain conditions, a particular (pure) sample of silver has melted at 1,100°C. We are confident if we heated another (pure) sample of silver, or re-heated our first sample, to that temperature, it would melt. More generally, we are pretty confident that *any* theory can be made to account for as many single observations as we like by the same experiment over and over again. Therefore, what tells us whether one theory is better than another is not the number of *single* observations it can explain, but the number of *types* of observations.

The above considerations suggest that, for the purposes of comparing competing scientific theories, what should count as a single component of data is a single pattern or regularity in our observations. A single component of data, on this suggestion, is a *class* of observations, all of which exemplify what we would regard as a single pattern.

Let us consider how this idea relates to our way of individuating components of theory. It has been suggested that we take a single component of theory to be an unrestricted universal generalization. Suppose it to be the sentence:

$$\text{For all } x, \text{ if } P(x) \text{ then } Q(x). \tag{3}$$

But suppose we have also made a series of observations of things that were P also being Q. This will count as a single component of data. Conditional (3) will therefore be a single component of theory that explains a single component of data. So for (3), the ratio of components of theory to components of data, and therefore its degree of independence from the data, will be one. More generally, on this view, simple empirical generalizations such as "All crows are black", "All silver melts at 1,100 degrees centigrade" and "All water at 1 atm boils at 100°C" will have a degree of independence of one.

The question naturally arises, "How, then, does a theory ever acquire a degree of independence greater than one?" One way – although not the only way – in which a theory can do this is by explaining a *number* of regularities in our observations by postulating a single, underlying regularity at the theoretical level. For example, we can explain a range of regularities at the observational level concerning the properties of heat, for example that it is caused by friction, that it causes things to expand, that it is caused by compression, that it is associated with "Brownian motion" and so on, by postulating that heat is the vibration of tiny particles elastically colliding with each other. Here a large number of regularities at the observational level are explained by postulating what are, plausibly, a smaller number at the theoretical level. A theory gets to be a "good" theory, or to have a degree of independence greater than one, by finding some way of explaining a number of distinct observational regularities in terms of a single or a smaller number of explaining regularities. One theory is better than another if it explains the same set of observational regularities by postulating fewer explaining regularities than the other.

EXPLAINING PARTICULAR STATES OF AFFAIRS

So far, our concern has been with the degrees of independence of explanations of an observational regularity or a collection of such regularities. But sometimes science gives us explanations of *particular* things: a mountain range, a system of rings around a planet, the solar system, a galaxy and so on. Let us begin by considering scientific explanations of the solar system. More specifically, let us consider the following two explanations:

> *Explanation 1* Originally there were nine nearly spherical planets orbiting the Sun in the same direction in nearly circular orbits in nearly the same plane and three non-spherical planets orbiting the Sun in the opposite direction. Passing meteors destroyed the three orbiting in the opposite direction, leaving the nine we now have. This system has continued to exist up to the present day, because the Sun and the nine planets obey Newton's three laws of motion and law of gravitation. The Sun has continued to emit warmth, because it is powered by thermonuclear fusion.
> *Explanation 2* Originally, there was a large cloud of dust with some angular momentum. The particles in the cloud were subject to Newton's laws of motion and law of gravitation. Gravity caused the cloud to contract, with most of the mass falling to the centre, and gravity, together with the angular momentum of the cloud, caused it to form into a spinning disk. Gravity also caused the central mass, and other inhomogeneities of mass throughout the disk to form spheres. Hence we now have a system of nine nearly spherical planets orbiting the Sun in more or less the same plane in nearly circular orbits. The mass of the central body was sufficient to initiate thermonuclear fusion.

Plainly, explanation 2 is good while explanation 1 is bad. Of course, we now have independent evidence that explanation 2 is (at least roughly) correct. But even if we did not have this additional evidence, we would still regard the first explanation as being a lot worse than the second. Why is this so? First, let us note that both explanations make use of exactly the same *laws*. One of the reasons why the first is such a poor explanation is because what we want explained is already there in the "initial conditions" of the explanation. The first explanation does not *reduce* the number of initial conditions that need to be explained. The second explanation, on the other hand, is able to explain many things – why nearly all the mass of the solar system is located at its centre, why the Sun and planets are all nearly spheres, why all the planets orbit in the same direction, and so on – simply by postulating a large cloud of gas with some angular momentum. So, intuitively, the second explanation has a much lower ratio of components of explaining theory to components of data than does the first explanation, and so it would appear to be much more independent of the data than the second explanation.

KOLMOGOROV–CHAITIN COMPLEXITY AND THE
INDEPENDENCE OF THEORY FROM DATA

The notion of the independence of theory from data can be distinguished from the notion of Kolmogorov–Chaitin complexity. Kolmogorov–Chaitin complexity can be explained as follows. Let G be a string of symbols. For definiteness, suppose G is 10101010101010 …. Then there will be ways of representing the string G that are shorter than G itself. One such representation will be "Write down '10' repeatedly." Let $S_L(G)$ be the *shortest* of all representations of G possible in language L. Here, the length of $S_L(G)$ is given simply by the *number of symbols* in $S_L(G)$, represented by $|S_L(G)|$. The number $|S_L(G)|$ is the Kolmogorov–Chaitin complexity of G in language L.

We can, perhaps, define the "Kolmogorov–Chaitin simplicity" of an expression as the inverse of its Kolmogorov–Chaitin complexity. Clearly, "Kolmogorov–Chaitin simplicity" bears some similarity to the notion of the independence of theory from data, and Kolmogorov–Chaitin complexity to the notion of dependence. But there are also a number of differences.

First, let us recall that there are two aspects to the notion of *dependence* of theory from data: the post hoc aspect and the complexity aspect. The notion of Kolmogorov–Chaitin complexity makes no reference to a post hoc aspect.

Second, in determining the degree of dependence (or independence), we must restrict ourselves to languages expressed in basic natural, or straight, predicates. No restriction of this sort is (explicitly) made in the definition of Kolmogorov–Chaitin complexity.

Third, the Kolmogorov–Chaitin complexity of a string is given by the number of symbols in a string. But the dependence of a theory is given by the number of

DECs it contains (more specifically, by the ratio of the DECs to the number of components of data explained). The relevant "unit" in Kolmogorov–Chaitin complexity is the symbol; in dependence it is the DEC. Moreover, for purposes of theory comparison, the type of DEC that usually plays the important role is (it will be argued) the unrestricted universal generalization of the form "$(x)(P(x) \rightarrow Q(x))$". That is, the basic unit that is counted when determining dependence is different from that counted when determining Kolmogorov–Chaitin complexity.

For our purposes – explaining the empirical successes of science – it is more appropriate to use the notion of independence of theory from data than the notion of Kolmogorov–Chaitin complexity.

The post hoc aspect of the independence of theory from data is required to deal with the "New Zealand sheep farming" objection and related difficulties.

The restriction to basic natural predicates is required to rule out theories expressed in terms of grue/bleen-type predicates and other artificial constructions. Unless some such restriction is made, it would be possible to re-express the most complicated, implausible or ad hoc theory in such a way that both its Kolmogorov–Chaitin complexity and the ratio of its DECs to components of data explained were very low. We can illustrate this with an example. Let $p_1, p_2, p_3, \ldots, p_n$ be the set of all space-time points ever observed, and let $Q_1, Q_2, \ldots, Q_n$ respectively be the observed properties of these space-time points. Let us now introduce the following predicate Q: A space-time point has Q if and only if that space-time point is p_1 and it has Q_1, or it is p_2 and has Q_2, and so on. Then the "theory" "Everything is Q" will entail all the empirical observations ever made. Moreover, if no restrictions are placed on the type of predicate that can be used, it will have a very low Kolmogorov–Chaitin complexity. But it is clearly a very bad theory, and neither does it intuitively give us what we would regard as "understanding".[6] Moreover, it will not count as a theory with a high degree of independence on the view offered here. When the predicate Q is defined in terms of basic natural predicates, it will turn out to have a very low degree of independence. It will turn out to be no better than a mere list of all our observations.

Our aim is to explain certain *predictive successes* of science. Clearly then, the theories with which we will be concerned, and which are to count as being good theories on the framework developed here, must be capable of *making predictions*. There is, however, no necessary reason why the briefest expression of a string of symbols (that is, the expression of the symbols with the lowest degree of Kolmogorov–Chaitin complexity) should also be *predictive* at all. It may say nothing more than is said by the string of symbols itself. An unrestricted universal generalization, on the other hand, is predictive. Moreover, in Chapter 3 it was argued that an unrestricted universal generalization that explains a body of data will have a higher probability than any other specific generalization from the data. So it is appropriate to use a framework which says theories expressed in terms of unrestricted universal generalizations are best.

THE RELATION BETWEEN THE APPROACH ADVOCATED HERE AND BAYES'S THEOREM

It is worth briefly discussing the relationship between the approach developed here and Bayes's Theorem. One version of Bayes's Theorem is as follows:

$$Pr(H, e) = \frac{Pr(e, H).Pr(H)}{Pr(e)}$$

where H is the hypothesis undergoing confirmation and e is the evidence adduced in support of H. Assuming that H logically entails e, the probability of e given H, that is Pr(e, H), will presumably be 1. Hence, under such circumstances, Pr(H, e) will be given by Pr(H)/Pr(e). So, if two hypotheses H and H* both have the same empirical consequences, H will be more likely than H* in the light of evidence e iff the prior probability Pr(H) is higher than the prior probability Pr(H*).

In this chapter it has been argued that if hypotheses H and H* have the same empirical consequences, but H is more independent from the data than H*, then H, for our purposes, is to be preferred. How does this comport with Bayes's Theorem? The first thing to note is that Bayes's Theorem and the view advocated here make claims about subtly but importantly different things. Bayes's Theorem tells us that, under the conditions described, the theory with the higher prior probability will, in light of the evidence, have the higher probability of being *true*. But on the view advocated here, the theory with the higher degree of independence from the data does not have a higher probability of being *true*; it merely has a higher probability of having true *empirical consequences*. Bayes's Theorem is about the truth of the *theory*; the view advocated here is about true *empirical consequences*. And so it might seem that there is no possibility of conflict between the approach advocated here and Bayes's Theorem.

However, there may be potential conflict. Clearly, if a hypothesis H is true, then all its empirical consequences will also be true. So, Bayes's Theorem tells us that if H has a higher prior probability than its rivals, then, under the appropriate conditions, its empirical consequences will also have a higher probability of being true than those of its rivals. But the view advocated here says it is not higher prior probability that leads to this, but higher independence from the data. The two views are therefore potentially in conflict.

Whether or not the two views really are in conflict will depend on the relation between the notions of "prior probability" and "independence from the data". Here the claim will not be defended that the independence of theory from data is a measure of the prior probability of a theory. However, in Chapter 5 it will be argued that the independence of theory from data is closely correlated with many features of scientific theories traditionally seen as virtues thereof. These include features such as simplicity, symmetry and, of course, lack of ad hocness. But it is just these features that are often seen as indicators of high prior probability in a theory. To that extent, the approach offered here would seem to at least cohere with the use of Bayes's Theorem as explicating scientific inference.

CONCLUDING REMARKS

In this chapter it has been suggested that one property that does at least many of the jobs to be done by property M is the *independence of theory from data*. An argument has been presented for the thesis that if a theory is highly independent of the data it explains, then it is *likely* that the data have a law-like tendency to conform to that theory.

It is also worth at this point summarizing where we are up to in our goal of producing an explanation that satisfies the four criteria of adequacy – as stated in Chapter 2 – for an explanation of the phenomena. To recap, the criteria of adequacy are as follows:

> *Criterion 1a* Property M must be *accessible*; more specifically, it must be more accessible than the forms of success of which it is taken to be an indicator (the accessibility requirement).
>
> *Criterion 1b* Property M must be such that we can explain why we have preferred theories with M, rather than any of the other many highly accessible properties of theories (the explicability requirement).
>
> *Criterion 2* An account which *merely* explains how we have managed to hit upon theories with property M would not be satisfactory; the account must also explain why it is that theories with property M (tend to) enjoy the forms of success exemplified in the three phenomena.
>
> *Criterion 3* Any satisfactory explanation of the phenomena must not leave it merely as a fortunate fluke that the type of theory that we have preferred also happens to be the very same type of theory as that which (tends to) enjoy the forms of success exemplified in the phenomena.
>
> *Criterion 4* Any satisfactory explanation must be able to account for the actual *historical examples* of the phenomena.

The prospects for the notion of independence of theory from data being able to meet criterion 1a – the accessibility requirement – seem promising. On the account offered here, the type of theory that is preferred and which (it will be argued) enjoys the forms of success exemplified in the phenomena is one that is highly independent of the data. A theory is highly independent of some body of data D iff each and every part of D exemplifies what we would intuitively regard as the same pattern, or a small number of such patterns. Therefore, in order to be able to determine whether or not a theory is highly independent of the data, all we need is a knowledge of the data and our intuitive notion of "the same pattern". The property of independence is a highly *accessible* property.

Criterion 1b is the requirement that we must be able to explain why scientists prefer property M *rather than* other accessible properties of theories. The strategy adopted here consists of three parts: (a) identify property M with the independence of theory from data, (b) argue that an *a priori* justification of the preferability of highly independent theories can be given and (c) argue that scientists have a (possibly implicit or inchoate) knowledge of this justification. In this chapter we have only attempted (b). We will undertake (c) in a later chapter. However, it is worth noting that the *prospects* for us being able to do this seem reasonably optimistic. The notion of independence is defined in terms of a basic natural predicate. We can identify such predicates by appealing to the ease with which their use can be taught. Moreover, our initial statement of the justification of independence, based on the sequence S1, was an explication of our intuitive reasons for preferring R3 to the other rules. So the definition of independence, and the justification of it, relies on pre-existing intuitions. Given that it is based on *pre-existing* intuitions, there does seem to be at least some possibility of showing that scientists have implicit or inchoate knowledge of the argument for independence given here.

Criterion 2 is that any satisfactory account must not only be able to explain why we actually prefer properties with M, but must also be able to explain why theories with M are successful in the ways exemplified in the phenomena. Although we have not explained any of the phenomena in this chapter, we have taken the first step towards the development of the explanations. It has been argued that if a theory is highly independent of the data, it is likely that there is a law-like tendency for the data to conform to that theory. Consequently, it is also more likely that the data would continue to conform to the theory if measurements were made in different locations or under different conditions. But, on the face of it, this would at most provide an explanation of *familiar* predictive success, since the predicted observations are just more instances of the pattern(s) already exemplified in the data, rather than instances of entirely novel patterns. So, at least on the face of it, it appears that the argument of this chapter would at most provide an explanation of familiar predictive success. But it will be argued in the next chapter that this is, in fact, not so: the argument of this chapter *does* provide us with an explanation of novel success.

Criterion 3 was that any satisfactory explanation of the phenomena must ensure that it is not merely a fortunate fluke that type of theory we prefer is also the type of theory that is successful. As we have already noted, the approach advocated here, if it is otherwise successful, can meet this criterion, because the fact which explains why scientists prefer theories that are highly independent of the data – their (implicit or inchoate) knowledge of the existence of a justification of the preferability of such theories – is the very same fact as that which will be used in the explanation of the success of those theories.

The fourth and final criterion of adequacy that must be met by an explanation of the phenomena is that it must be able to explain actual historical cases of success. This task will be undertaken in later chapters.

CHAPTER 5

Some more success-conducive properties of theories

INTRODUCTORY REMARKS

In the previous chapter, the notion of the independence of theory from data was introduced and defined. This notion was intended as a *partial* measure of the lack of ad hoc dependence of a theory on the data it explains. It was also argued that there is reason to believe this notion will help us to explain the phenomena with which we are here concerned. And in later chapters it is indeed the notion of independence that plays the most prominent role in the explanation of the phenomena. But, as we will see in this chapter, the notion of independence, as defined, is not by itself *sufficient* to account for all cases of the phenomena: some more notions are needed. Neither is it, by itself, a complete account of the features of a theory that make it more likely to enjoy subsequent empirical success. The aim of this chapter is to describe and justify more of these success-conducive properties of scientific theories.

Although this chapter describes a range of different properties of theories, they all have something in common: they all indicate that a theory with these properties has an increased chance of future empirical success. Moreover, for all these properties, the reasoning justifying this conclusion is the same: it is highly unlikely that it should merely be due to chance that the data should be explainable by a theory with these properties, so it is likely not due to chance that the data is so explainable, so it is reasonable to believe there is a propensity for the data to be so explainable and that it will continue to be so explainable when observations are made of other locations within the data.

The question naturally arises: do the properties dealt with in this chapter provide an exhaustive list of all the properties of theories conducive to future empirical success? No claim is made here that this list is comprehensive: in fact, it is not. It is only claimed that the properties described in this chapter – and in the previous two – are sufficient to explain the historical examples of success considered in the next three chapters.

INTRA-DEC INDEPENDENCE

The independence of a theory from data was defined as the ratio of components of data explained by the theory to its dependent explanatory components (DECs). It plainly follows from this that one theory is more independent than another if it has a higher ratio of components of theory explained to DECs. But if this is to justify the claim that a more independent theory is *more likely to be empirically successful* than a less independent theory, it must be assumed that the DECs *within* the competing theories are equally good, or equally likely to lead to correct predictions. But, as we will see, this need not be the case. The notion of independence, *as so far defined*, only takes us a part of the way towards a complete account of lack of ad hoc dependence on the data. It also only takes us a part of the way in developing a full account of the features of a theory that are responsible for its subsequent empirical success. One of the things we also need is an account of the degree of independence of *individual* DECs. One aim of this chapter is to develop such an account.

It is worth clarifying the terminology that will be used throughout the remainder of this book. As we noted above, the notion that does the bulk of the work in explaining the phenomena with which we will be concerned is the notion of independence described in the previous chapter, that is the type of independence that is a property of *systems* of DECs. So, in what follows, we will simply refer to this notion as "the independence of theory from data". But in certain contexts we will also need to use a notion of independence that applies to individual DECs. We will refer to this notion as "intra-DEC independence". But, unless otherwise specified, when we speak of "the independence of theory from data" we will be referring to the property of *systems* of DECs.

COMPARING THE INDEPENDENCE OF DECS

So far, we have assumed that all components of theory are equally likely to lead to true predictions. But in some cases this is very implausible. Suppose we have two variables, x and y, and we determine the value of y for values of $x = 0$, 1, 2 and 3. We obtain the following results:

x	y
0	2
1	3
2	4
3	5

We note that there are (at least) three rules that can account for the observed values of x and y (Hempel 1966: 41). These are as follows:

1. For all values of x, and for all values of y, the value of y is given by $x + 2$.
2. For all values of x, and for all values of y, the value of y is given by $x^4 - 6x^3 + 11x^2 - 5x + 2$.
3. For all values of x, and for all values of y, the value of y is given by $x^5 - 4x^4 - x^3 + 16x^2 - 11x + 2$.

It seems plain that rule 1 is preferable to the others. But, since all three rules are unrestricted universal generalizations, they should all count as single components of data. So, on the account developed so far, all three rules would be equally likely to lead to correct values of y for other locations of the data, specifically for values of x greater than 3. But this certainly does not square with our intuitions in this case.

However, it is clear that within the general framework developed so far, good reason exists to prefer the first rule to the others. Recall that on the view advocated here, the fundamental inference used in considering whether a hypothesis is likely to be correct is the following: "Is it likely that the data could have conformed to this hypothesis by chance? If not, it is likely that there exists a propensity for the data to conform to this hypothesis, and so it is likely that at other locations the data will continue to conform to this hypothesis." And it is not too difficult to see that a preference for the first rule can be justified using this inference.

In at least rough outline, the argument is quite straightforward. It seems unlikely that it could be due to chance that the data should conform to a rule as simple as rule 1: the vast majority of quadruples of ordered pairs of numbers could not be explained by a rule so simple. Therefore, by reasoning already defended, it is likely that there is a propensity for it to conform to that rule. But it seems rather more likely that there should be some highly complex rules, even universally quantified rules, that could account for the data. No matter what the four ordered pairs had been like, it seems plausible that there should be *some* highly complex universally quantified rule that could generate them. Therefore, in the case of rules 2 and 3, the inference to the conclusion that it could not be due to chance that the data exemplifies those rules is less compelling than it is with rule 1. We therefore have reason to prefer the hypothesis that there is a propensity for the data to conform to rule 1 to the alternative hypotheses that there is a propensity for it to conform to rule 2 or to rule 3.

In the above paragraph it was suggested that it seems more likely that the data should exemplify some rules that are as *complex* as rules 2 and 3. So the considerations of the above paragraph give us reason to prefer unrestricted universal generalizations – that is, components of theory – that are simple rather than complex. But this, of course, raises the question "What do we mean by the terms 'simple' and 'complex'?" The problem of defining the notion of complexity is a large and ongoing problem in the philosophy of science. However, the argument of the previous section helps us to get a little clearer on the notion of complexity that is relevant in the present context. Why do we feel that it is easier to find some rule *like* rules 2 and 3 that exemplify the data than it is with rule 1? The answer, surely, is because there are *so many* that are "like" rules 2 and 3, while there are much fewer "like" rule 1. Consider the following two classes of equations:

1. $y = Ax + B$
2. $y = Ax^4 + Bx^3 + Cx^2 + Dx + E$

where, let us assume, A, B, C, D and E can each be any whole number between −20 and +20. Then the number of equations in class 1 will be 1,600, while the number in Class Two will be 102,400,000. The *a priori* probability that some rule capable of explaining any arbitrary body of data will be found in a class consisting of 102,400,000 members is a lot higher than the *a priori* probability of there being such a rule in a class with only 1,600 members. So the *a priori* probability that, no matter what the data had been like, there would have been *some* rule with the same number of parameters as rule 2 that explained the data is much higher than the *a priori* probability that there should have been a rule with the same number of parameters as rule 1.[1] Hence, in the case of rule 2, we can say it is more likely to be due to chance that the data has a propensity to conform to that rule. More generally, we may say that if two DECs both explain the same body of data, but one has fewer parameters capable of taking on any value than the other, while they are otherwise equal, we have good reason to prefer the DEC with fewer parameters. In what follows we will refer to this as the argument for the *parsimony of parameters*. Note that it falls short of being a *complete* account of intra-DEC independence, since it stipulates the DECs to be compared must be "otherwise equal".

JEFFERYS AND BERGER ON OCKHAM'S RAZOR

William Jefferys and James Berger have published influential work within a Bayesian framework that in some respects resembles the argument of the previous section (Jefferys & Berger 1992: 64–72). Jefferys and Berger seek to justify our preference for simple theories. Moreover, they argue that it is a feature of simple theories that their agreement with observation is much less likely to be due to chance, and this gives them a higher likelihood of being correct than less simple theories. Clearly then, the views of Jefferys and Berger on the face of it appear to be similar to those developed here. However, it will be argued that they are in fact in a number of respects different, and that the position of Jefferys and Berger fails to explain the phenomena with which we are here concerned.

The core of Jefferys and Berger's argument is as follows: if a theory is complex, it will have many "free parameters". But the larger the number of free parameters a theory has, the easier it will be to find a way to get the theory to fit the data, no matter what the data turn out to be. A complex theory, in their terminology, does not make "sharp" predictions. So the ability of a complex theory (i.e. a theory with a large number of free parameters) to fit the data does not provide impressive confirmation of the theory, and neither does it greatly increase the probability of the theory. A simple theory, on the other hand, has a smaller number of free parameters. There is a much lower chance that it will be possible to make a simple theory fit the data. In this sense, a simple theory makes much sharper predictions, and it is much more impressive confirmation of the theory if it turns out to be able to explain the data. Consequently a simple theory is made more probable by its ability to explain the data than a complex theory.

The views of Jefferys and Berger closely resemble the argument developed in the previous section. Indeed, they are perhaps making the same point but with different terminology. First, let us note that a "theory" with free parameters is not, strictly speaking, an *explanation* of anything. It might perhaps be called an "explanation schema", or a class of explanations. We only have an explanation of some data when the parameters are assigned specific values in such a way that they theory logically implies the data. So a "theory" with free parameters can be regarded as a class of possible explanations that all have the same form and which all have, in our terminology, the same degree of independence from the data. If the theory has a large number of free parameters, then the corresponding class of potential explanations will all have a low degree of independence from the data, so it will not be surprising if some member of that class can actually explain the data. But if a theory is simple and has few free parameters, the corresponding class of potential explanations will be much smaller, so it will be much less likely merely due to chance that one of them would be able to actually explain the data. The argument of the previous section and that of Jefferys and Berger would, that is, seem to be saying the same thing with different words.

However, it is clear that, as it stands, the position of Jefferys and Berger is subject to difficulties we addressed in Chapter 4. Unless some restriction is placed on the type of predicate that can appear in a theory, it will always be possible to find a very simple theory, or theory with few free parameters, that can explain any body of data. Suppose, for example, some finite body of data has been obtained relating two magnitudes X and Y. When values of Y are obtained for X = 1, X = 2, X = 3 and so on, they vary in a completely random manner. For example, when X = 1, perhaps Y = 54, when X = 2, Y = 7584, when X = 3, Y = −87, and so on in a random way. We then introduce a predicate R that applies to ordered pairs. An ordered pair has R iff it is <1, 54> or <2, 7584> or <3, −97>, and so on for all values in the data set. Then the "theory" "The data conforms to R" will have no free parameters, and will be very simple. But it is obviously a very bad theory, and would be of no use for making novel predictions. So the apparatus developed by Jefferys and Berger, despite its similarity to that developed here, is not sufficient for the tasks with which we are here concerned.

There are, of course, other difficulties with the account of Jefferys and Berger, in so far as it purports to be a general account of why we prefer simple theories. Perhaps most obviously, their account *only* applies to pairs of theories both of which have free parameters. But we can easily imagine a situation in which we have two theories neither of which has any free parameters, both of which make perfectly "sharp" predictions, but where one is much simpler than the other. A case of this type might arise when we are comparing two possible lines, both of which pass through all of the obtained data points. If one of those lines were a straight line and the other a line that snaked around, we would clearly prefer the straight line on the grounds that it was simpler. Yet since both lines would precisely predict future values (at least within a limited range), neither need correspond to equations with any free parameters at all.

Finally, the account of Jefferys and Berger would only appear to apply to those types of theories that could, with some plausibility, be capable of having "free parameters". But not all explanatory theories are like this. In particular, theories expressed in natural language rather than the language of mathematics may be difficult to represent in this way. Yet here, too, it seems natural to prefer the simpler theory.[2]

THE INDEPENDENCE OF SYSTEMS OF DECS AND OF INDIVIDUAL DECS

We have now developed accounts of two features of theories that are indicators of their likely correctness. These are the independence of theory from data, or the ratio of components of data to components of theory, and the parsimony of parameters, which is an incomplete measure of intra-DEC independence. Plainly, then, the best theories will be those that maximize independence in both of these senses.

At this point it might seem natural to ask the question "How are these two features related? In particular, if theory T1 has fewer DECs but they posses a lower degree of intra-DEC independence, while T2 has more DECs but they have a lower degree of intra-DEC independence, which of the two theories is, all things considered, the better?" However, given the overall aims of this book, it is not necessary to consider this question. Recall that the aim of this book is merely to explain certain forms of scientific success. It is not to provide some comprehensive set of rules of method that would enable a scientist, in any possible situation, to select the best theory from among some set of alternatives. In order to provide an explanation of success, it is sufficient to show that the successful theories are highly independent in both ways. Moreover, in Chapters 6–8, explanations of this kind are developed and applied to cases from the history of science.

THE LOWEST POSSIBLE DEGREE OF AD HOCNESS: CONSERVATION LAWS

We have noted that one way in which some DECs have lower degree of (intra-DEC) dependence than others. The question naturally arises: "Is there some class of DECs the members of which possess the lowest possible degree of ad hocness?" In this section it will be argued that there are.

The notion of independence, whether of systems of DECs or of individual DECs, is intended as a measure of the extent to which the inference to the conclusion that it could not be due to chance that the data has a propensity to conform to a particular theory is a justified one. If a theory has a low degree of independence, or many ad hoc features, the inference to that conclusion is weak, because it is likely that, no matter what the data had been like, there would have been *some* theory with that many ad hoc features that would have been able to explain it. We therefore cannot in such a case make the inference to the conclusion that it could not be due to chance that the data should exemplify this particular theory. But if a theory exhibits a *low* degree of ad hocness, that is if it has very few features dependent on the data, then the inference to the conclusion that it is not just due to chance that the data exemplifies such a theory is a *stronger* inference. This type of inference will be at its strongest when there is some theory that, in some sense, possesses the *smallest possible* number of ad hoc dependent features. What theory or class of theories posses this characteristic? Here it will be argued that it is those theories that say some property is *conserved*. Consider the following sequence:

$$2, 2, 2, 2, 2, 2, 2, \ldots \tag{1}$$

One rule that generates this body of data is "Write '2' in every position." We will call this "rule (a)". On the face of it, this rule has a very low degree of ad hocness. It has only a single DEC which itself has a low degree of intra-DEC dependence. Moreover, at an intuitive level, it seems to be about as simple as a rule for generating an indefinitely large sequence can possibly be. Intuitively, there is perhaps only one other category of rules that might be thought to posses so few dependent features, and these are the rules that generate bodies of data like:

$$1, 2, 3, 4, 5, 6, 7, \ldots \tag{2}$$

A rule that generates this body of data might be "Write down the natural numbers." We will call this "rule (b)". On the face of it, this rule (b) might seem to have as few, or even fewer, dependent features than rule (a). But it will be argued that this is not the case: rule (a) has fewer dependent features.

Let us begin by considering why it might initially seem that it is rule (a) that is more dependent. It might be thought that there is one dependent feature of rule (a) that is not present in rule (b), and that is that rule (a) specifies that we must write down the number "2" rather than some other number, whereas rule (b) does not have any such arbitrary feature. But a little reflection is sufficient to show that this arbitrary feature is also present in rule (b). Suppose that the numbers in (1) are the results of measurements of some property. That the number that appears in (1) should be "2", rather than say, "4" or "7" or "23", is, of course, a result of our particular choice of units. With a different choice of units, the sequence: "23, 23, 23, 23, ..." would, of course, register the very same pattern in nature. However, this same feature is to be found in (2). Suppose the numbers in (2) are measurements of a property. The *difference* between each measurement is in each case 1 precisely. But that it should be equal to 1 is, again, plainly a result of our particular choice of units. What is in fact the very same pattern in nature could be registered by 2, 4, 6, 8, ... or by 10, 20, 30, 40, ... and so on. So this element of arbitrariness found in rule (a) (write down "2" rather than any other number) is also present in rule (b).

There is, however, an element of arbitrariness present in rule (b) that is *not* present in rule (a). Each element of (2) is obtained from the earlier element by the operation of *adding*. But it is an arbitrary feature of (2) that each element in it should have been obtained by adding rather than, say, subtracting or multiplying or dividing or performing some other operation. This element of arbitrariness is not present in (1), since each element can be obtained from the earlier element by doing nothing at all to it. Each mathematical operation (adding, multiplying, squaring, etc. has its inverse (subtracting, dividing, extracting a square root, etc.) But the operation of leaving a number as it is has no "inverse", or it is its own "inverse". So it is a dependent feature of any rule requiring us to perform an operation on a number that it stipulates that operation rather than its inverse. But the "operation" of leaving each number as it is is unique in not having that dependent feature. So rule (a) has fewer dependent features than rule (b). And so, we may conclude, by the remarks given in Chapter 4 (pages 60–64), that rule (a) has fewer dependent components than rule (b). But since there appear to be no possible rules that could posses fewer dependent components or aspects than either rule (a) or rule (b), we may conclude that rule (a) has fewer dependent components or aspects than any other rule. And, of course, the same can be said of other rules of the same form, such as "Write down '5' in every position", "Write down '38' in every position" and so on.

Now, let us again suppose that a sequence such as (1) is the result of a series of measurements of some property. More particularly, we will suppose that it is a series of readings of measurements of some property in a *closed* system. Plainly, this sequence of readings would correspond to the *conservation* of the property being measured. Rules like (a) generate the same observation claims as would be generated by a true conservation law. So we may say that *conservation laws* postulate those patterns in the data that have the minimum possible degree of dependence on the data. In this sense, conservation laws have a special status. Provided that a conservation law does explain a body of data, the foregoing considerations show it has the highest possible degree of intra-DEC independence that it is possible for an empirical claim to have.

THE CLOSE-TO-*A PRIORI* STATUS OF CONSERVATION LAWS

Perhaps the most puzzling of all the phenomena with which we are here concerned is the fact that there are some theories which seem to be in some sense close to *a priori*, but which enjoy a surprising degree of empirical success. In this section it will be argued that the ideas developed so far suggest a number of explanations for the success of *a prioristic* theories.

As far as I am aware, all cases of *a prioristic* but surprisingly successful theories involve conservation laws, or are plausibly a consequence of conservation laws.[3] We have already noted that inverse square laws of force are precisely what we would expect to be the case if we conceive of force as being due to something emanating from the source of the force and which is *conserved* as it moves away from that source. In this section our concern will, therefore, be with the special status of conservation laws. It will be argued that it is a natural consequence of the views advocated here that simple conservation laws should be very close to *a priori*.

On the view advocated in this book, a theory is good if it has a high degree of independence from the data it explains. A theory has a high degree of independence from data if it explains a large number of regularities in terms of a small number of regularities. The independence of a theory from data is the *ratio* of components of data explained to *dependent* explanatory components of theory: the larger the number of components of data explained, and the smaller the number of *dependent* explanatory components of theory, the better the theory. A component of theory is *dependent* on the data if it is postulated after observing the data, and the reason for asserting is that it can account for one or more components of data.

Let us now suppose, hypothetically, that there exists a genuinely *a priori* theory which is able to explain at least one empirical regularity. Since it can explain at least one empirical regularity, there will be some components of theory it can explain. But if the theory is genuinely knowable *a priori*, there will be a reason for asserting it that is independent of any empirical data it may explain. An *a priori* theory will not be dependent on the data. Therefore, an *a priori* theory will have *no dependent* explanatory components of theory. So if there were a genuinely *a priori* theory that could explain some data, the ratio of its components of data to dependent explanatory components of theory, and hence its degree of independence from the data, would be infinitely large. It would, on the view advocated here, be the best possible type of theory.

Recall the discussion of conservation laws given in the previous section. There it was argued that conservation laws posses a higher degree of intra-DEC independence than any other empirical claim. They are, therefore, the type of DEC for which the inference to the conclusion that the data will conform to them is at its strongest. Now let us consider the range of phenomena *explained* by conservation laws. In the world around us, we continually see evidence that is capable of being explained by the law of conservation of mass or substance: when we pour milk from a carton into a glass, the total amount of milk seems, as far as we can tell, to be pretty much unchanged; when we dig some earth from the ground, the amount of earth in the heap seems to be about the same as the volume of the hole we have created, and so on. Of course, as we noted in Chapter 2, the claim that mass is *precisely* conserved is certainly not the only claim capable of accounting for these observations, but it surely seems to be among those that can. And since just about everything in our environment has mass or is made of matter, we are continually confronted with observations that can be accounted for by the claim that mass or matter is conserved. Apparent conformations of this idea are ubiquitous: everything that happens in the familiar world around us can, at least at the level of accuracy of observation available to common practice, be seen as confirming to this law. So, it would seem intuitively that the range of data that can be accounted for by the idea that matter is conserved comes pretty close to being as wide as it is possible for a body of data to be.

We have noted that the claim that matter is conserved possesses a higher degree of intra-DEC independence than any others capable of explaining the same phenomena. Moreover, at an intuitive level at least, the range of occurrences we see that can be seen as conforming to that law seems to come pretty close to being as broad as possible. So, again at an intuitive level, it seems that the idea that mass, or substance, should be conserved comes very close to having the highest possible degree of independence from the data. But, we noted above that if there were a genuinely *a priori* claim capable of explaining some data, it would have the highest possible degree of independence from the data. Hence, the law that mass, or matter, is conserved closely resembles an *a priori* statement: it has a degree of independence very like that which would be possessed by an *a priori* statement that also had explanatory power. This provides us with an explanation of why it is that, at an intuitive level, the idea that substance is conserved seems to be like an *a priori* statement. It also explains why the law of the conservation of mass continues to receive confirmation when we move from everyday observations to, for example, the chemical laboratory: by the argument given in the previous chapter (pages 67–9), the hypothesis that the data obtained by more accurate measurements will continue to conform to a conservation law has a higher *a priori* probability than any alternative hypothesis. Consequently, we have an explanation of why the law of conservation of mass should intuitively seem to be close to *a priori*, but nevertheless enjoy surprising empirical success.

The account given above of the *a prioristic* status of the law of conservation of mass appeals to the ubiquity, in our everyday experience, of apparent confirmations of that law. So, it could also, perhaps – but with somewhat less persuasive force – account for the *a prioristic* nature of some law such as the conservation of kinetic energy, or the conservation of momentum. But there are some other, more exotic conservation laws – such as those for spin or lepton number – which we obviously do not find confirmed in our everyday experience. So the account given here provides no explanation of why those more exotic laws should seem to be *a priori*. But this is surely exactly as it should be: surely these more exotic laws do *not* seem to be *a prioristic* in the way that something like the law of conservation of mass or momentum do seem to be.

This view provides an explanation of why the idea that *substance* will be conserved should seem to be close to *a priori*. So if, for example, electric charge is conceived of as a substance, it provides an explanation of why the law of conservation of charge should seem to be close to *a priori*. And, as we have already noted, it also provides us with an explanation of why certain inverse square laws of force should seem to be close to *a priori*: a law of force will be inverse square if the force is due to some substance emanating from the source that is conserved as it moves away from that source. An explanation of the *success* of conservation theories also follows directly. Laws saying that substance is conserved will have, it has been argued, the highest degree of independence from the data of any empirical claim. But it has already been argued that assertions with this feature are likely to enjoy predictive success in conditions other than those in which they were initially confirmed. Similarly, inverse square laws of force will also have a high degree of independence from the data, and so their empirical success can also be (probabilistically) explained. So we have an explanation of why at least *some* of those laws that seem to us to be like *a priori* truths also enjoy a surprising degree of empirical success.

THE SPECIAL ROLE OF LOW WHOLE NUMBERS IN SCIENCE

It is a fact that sometimes scientists quite readily conclude that since the measured value of some quantity is very close to a low whole number, the true value of the quantity *is* that low whole number *precisely*. We have already encountered the case of Coulomb, who established the correctness of the inverse square law of electrical force to within an accuracy of three per cent. But Coulomb did not assume that the value of the exponent was 2.005, or 2.02; he said it was 2 *precisely*. The assumption that if the measured value of some quantity is close to a low whole number, then the correct value is that low whole number exactly is perhaps most often made when assigning an exponent to the way a force diminishes over distance. But, plausibly, it occurs in other cases, too. For example, since observations led to the prediction that the neutrino ought to have an undetectably small mass, it was hypothesized that its mass was in fact zero. It has also been hypothesized that since the ratio of particles to anti-particles in the universe seems to be very close to one, its true value is exactly one. Mendel found that the ratios of certain heritable characteristics in populations was very close to 1 to 3, and conjectured that the true value was exactly 1 to 3. It might be felt that this is an unempirical and unjustified tendency towards oversimplification, but we should not overlook the fact that this tendency has led to some surprising predictive successes in physics and genetics, a theme we explore in detail in later chapters.

Let us begin by spelling out an argument for the preferability of low, whole numbers that is of the same form as the arguments developed earlier for independence. A natural way of stating the argument is as follows:

1. Suppose it to be the case that a measurement of some quantity Q has been made, and it has been found that the measured value lies very close to some low, whole number, and the *a priori* probability that the measured value should lie so close to a low, whole number is very low.
2. So it is likely that it is not due merely to chance that the number should lie so close to a low, whole number.
3. Therefore it is likely that there is a propensity for measurements of the quantity to lie close to a low, whole number.
4. Hence it is likely that future measurements of the quantity will lie close to the low, whole number.

The above argument is plainly of the same form as arguments defended earlier, and, plausibly, it does succeed in making a good case for the claim that future measurements of the quantity will also lie *close* to the low, whole number. But what it does *not* succeed in showing is that the true value of the quantity should be the low, whole number *precisely*. For this, another argument is needed.

There is one feature of our tendency to prefer low whole numbers that any justification of this practice must be able to explain. It is only *sometimes* that we prefer low, whole numbers. Coulomb preferred the hypothesis that the real value of the exponent was two to the hypothesis that it was, say, 2.02. But suppose we found a rock lying on a hillside, and measured its weight to be 2.02 kilograms. Even if the accuracy of our scales did not rule out the possibility that its mass was 2 kilograms precisely, we would have no temptation to assume the real value of the mass of this rock was 2 kilograms precisely. Any argument in support of the preferability of low, whole numbers ought to be able to account for the fact that it is only in *some* circumstances that we do prefer them.

We can get a clearer picture of what is going on here by contrasting a case in which we would conclude that a low, whole number is the correct value with one in which we would not. Suppose some natural property has been measured, and the obtained value is 1.9998. Suppose also that given the accuracy of the measurement techniques available, the possibility is not ruled out that the true value of the magnitude of the property is 2 *precisely*. Would we conclude that the true value was indeed 2 precisely? It depends. Let us assume that the obtained number is a ratio, and it is the ratio of the masses of two types of sub-atomic particles we will call Xons and Yons. In such a case I think we would very likely conclude that the real value was 2 precisely. Now let us consider an example of a very different kind. Suppose that the same number 1.9998 was the ratio of our obtained estimates of the number of sheep in Yorkshire and the number of elephants in Namibia. In this case I do *not* think we would be at all inclined to assume that the real value of the ratio of the two populations was 2 exactly. And I think we are not inclined to say this because there is *no reason* why it ought to be 2; more particularly, there is no reason why there ought to be any connection between the two populations that would bring it about that one would be twice the other. So if the number of Yorkshire sheep were exactly, or very nearly exactly, twice the number of Namibian elephants, we would have no hesitation in saying that it was merely a fluke that this should be so.

This can be contrasted with the imaginary example of the Xons and Yons. In this case we would be more inclined to draw the conclusion that the real value of the ratio was 2 precisely. And in this case we are also more ready to accept that there should be some reason why the ratio should be 2 precisely. For example, if the mass of an Xon is, to within the limits of accuracy of our techniques of measurement, twice that of a Yon, the hypothesis immediately suggests itself that an Xon is made up out of two Yons. And if this hypothesis is correct, then – disregarding the possibility that some mass might be converted into energy in the formation of an Xon – we would expect the mass of an Xon to be *precisely* twice that of a Yon. It is important to note that this inference still may have considerable force even if we do not *know* that an Xon is made up of two Yons. But, to the extent that it is a *plausible* hypothesis that an Xon is composed of two Yons, it is also reasonable to make the inference to the conclusion that the real value of ratio of their masses is exactly equal to 2.

In view of the preceding considerations, the following, more general hypotheses suggest themselves: Suppose it to be the case that the measured value m of some quantity is close to some low whole number n, the *a priori* probability that m should lie so close to n is very low, and the accuracy of our methods of measurement do not rule out the possibility that the correct value of the quantity should be n precisely. Then:

1. If it is a consequence of our general background beliefs that it should be just a fluke that m should lie so close to n, then we would not be inclined to say that the real value of the quantity was n precisely.
2. If it is a consequence of our general background beliefs that it may not be just a fluke that m lies so close to n, there must be some possible explanation, or class of explanations, of why m should lie so close to n. And if it is a consequence of that explanation that the value of the quantity should be precisely n, then we are inclined to accept that the correct value of the quantity is indeed n precisely.

That (1) and (2) seem to be on the right track is supported by the following imaginary case. Suppose we detect, coming from outer space, two very regular sources or pulses, which presumably come from "pulsar" stars. Call the two sources A and B. We measure the time intervals between the pulses coming from A and compare them with the intervals coming from B. We find that the intervals between the pulses coming from A and those coming from B are very nearly the same, that is the ratio of the intervals is very nearly one. Assume also that it is *a priori* highly unlikely that the ratio should lie so close to 1. Given the accuracy of our measurements, the possibility is not ruled out that the ratio is precisely equal to one. Would we in this case assume that the true value of the ratio was indeed precisely 1? Here we need to distinguish between two types of case: (i) the two sources come from very nearly the same part of the sky, and (ii) the two sources come from completely different parts of the sky. In the first case I think we would be inclined to say that the real value of the ratio was 1 precisely, but in the second case we would not. And this accords with the hypotheses suggested above. If the two sources came from very nearly the same part of the sky, and it was *a priori* unlikely that the intervals between the two sets of pulses should be so nearly the same, the hypothesis would suggest itself that the two signals actually had their origin in the very same pulsar star, and they appeared to observers on Earth as two distinct signals due to "gravitational lensing".[4] But it is a consequence of this hypothesis that, since the two signals come from the very same source, the time intervals between them should be exactly the same, and hence that their ratio should be precisely 1. On the other hand, if the two signals came from very different parts of the sky, then there would seem to be no explanation of why the ratio of the time intervals should be so close to 1, and so in this case we would say it was just a fluke that the ratio was so close to 1. Since there would be no explanation of why the two pulse intervals were so close to being the same, there would be no explanation entailing that the intervals should be exactly the same. And so in this case we would have no reason to believe that the ratio was precisely equal to 1.

LOW WHOLE NUMBERS AND LARGE WHOLE NUMBERS

The view suggested here explains why it is that we are, at least sometimes, inclined to accept that the real value of some quantity is a *low* whole number, but also why we are much less inclined to accept that the real value of a quantity should be some *large* whole number. Intuitively, while the *a priori* probability that the measured value of some measured quantity should lie close to some *low* whole number is low, the *a priori* probability that it should lie close to a very large whole number seems somewhat higher. If the measured value of some quantity was 0.9998, we may be tempted to suspect the true value was 1 precisely, but if the measured value of a quantity was, say, 4,792,085.9998, we would, I think, be rather less inclined to suspect that the true value was 4,792,086 precisely. *One* reason for this is because, while we may easily be able to imagine plausible mechanisms that would imply that the true value of a quantity should be one or two precisely, mechanisms that imply that the true value of a quantity should be exactly equal to some number such as 3,492,086 seem rather rarer or more likely to be ad hoc. Also, of course, there is a clear sense in which the *a priori* probability that the measured value should lie very close to a *low* whole number is much lower than the *a priori* probability that it should lie close to a much larger number.[5] And so the inference to the conclusion that it could not be due to chance that the measured value lies close to a whole number is much stronger in the case of a low number than it is in the case of a large number.

Perhaps the most obvious type of case in science in which we are inclined to assume that the real value of some quantity is a low whole number are "inverse square" force laws, such as the law of gravitation and Coulomb's law. That scientists should have been inclined to accept that the precise value of the exponent is the inverse of the *square* of the distance is easily explainable on the view suggested here. Suppose we find that the intensity of some force decreases with the distance from its source, and that it does so in a way that is at least approximately in keeping with an inverse square law. It is natural suppose that the force is due to something emanating from its source, and that the force becomes weaker as we move away from the source because that which is emanating becomes more rarefied or dilute. But if we suppose that the total quantity of that which is emanating from the source is conserved as it moves away, it follows that the law describing how the intensity of the force diminishes with distance must be an inverse square law. That is, it is a consequence of what is surely the most (initially) plausible mechanism concerning how the force is propagated that the value of the exponent should be *precisely* 2.

Does this give us good reason to accept that the correct value of the exponent is, indeed, 2 precisely? At first it might be thought that this depends on how plausible it is that the "substance" emanating from the source of the force should be *conserved* as it moves away. And in the previous section it was argued that that such a hypothesis is as lacking in ad hoc dependence as it possible for a hypothesis to be. And so it is a hypothesis that we can expect to have future empirical success. But in the next section it is argued that, even if we did not have this reason to prefer such hypotheses, there may still be good grounds for preferring hypotheses that use low whole numbers.

THE AIM INFERENCE

It is a familiar fact that our confidence in some claim is increased if two independent sources of information make the same claim, and the *a priori* probability that they should agree is rather low. For example, if two quite different ways of measuring the height of a mountain independently tell us that the height of the mountain is 25,807 feet, we are more confident that the height of the mountain is indeed 25,807 feet than if we had had only a single source of information. Or again, if two witnesses independently tell us that a man wearing tartan trousers and a yellow cloak and carrying a violin case was seen running from the scene of the crime, we are thereby made more confident in the accuracy of this improbable description than we would be if there had been only one witness. This form of confirmation has been aptly called "cross-bracing".

There is one aspect of cross-bracing with which we will here be concerned. Suppose we had no prior knowledge of the reliability of either of the witnesses who reported the strangely dressed figure fleeing the crime scene. They might for all we know be perfectly reliable, but they might be fantasists or delusionals or habitual liars. But if they both independently give us such an *a priori* improbable report, not only is our confidence thereby increased that what they said was true, but our confidence is also increased in the reliability of the witnesses themselves. Or again, suppose we believed the two methods used to measure the height of the mountain might both be prone to error. If both methods were in close agreement, and the closeness of this agreement was *a priori* highly improbable, then our confidence in the reliability of both the methods would be increased. More generally, we may say, the *a priori* improbable agreement of independent methods permits us not only to have more confidence in the correctness of the result upon which they agree, but also to have more confidence in the methods themselves. We will call this inference from the agreement of independent methods to the greater reliability of those methods the AIM inference. It is an inference we will use later in explaining the phenomena.

An argument can be given for the AIM inference that conforms to the general type of inference we have been using. We can represent this argument as follows:

1. Suppose two independent methods M1 and M2 for measuring a quantity Q have given us exactly, or very nearly exactly, the same result, and the *a priori* probability that they should be in such close agreement is very low.
2. Plainly, either the result given by the two methods is (nearly) accurate or it is not. If it is not, then two distinct sources of error happen to have led us to (nearly) the same result. But this is *a priori* unlikely.
3. Therefore the result given by the two methods probably is (nearly) accurate.
4. Hence the result given by one of the methods, such as M1, is probably (nearly) accurate.
5. But it is unlikely that M1 should have given us a (nearly) accurate result by chance.
6. Therefore, it is likely that there is a propensity for M1 to give us accurate results. The same reasoning also shows it is likely there is a propensity for M2 to give us nearly accurate results.

Perhaps the most controversial claim in this argument is step 2. But closer inspection, I think, shows step 2 to be quite unexceptionable. Let us return to our example of measuring the height of a mountain by two independent techniques. Suppose one technique uses theodolites while the other uses climbers equipped with measuring rods, who measure the height of the mountain as they climb it. They agree that the height of the mountain is 25,807 feet. How might we explain their agreement, while at the same time holding that both methods are substantially in error? We can say that sources of error are, roughly, of two sorts: errors that come from mistakes in the underlying principles used (errors of principle) and errors in the application of details (errors of detail). An error in the underlying principles used by the theodolite method might be, for example, use of the wrong theorems of trigonometry in calculating the height of the mountain. An error of principle in the method using measuring rods might be, for example, the mistaken use of a rod that was in fact *not* a foot-rod. An error of detail in the theodolite method might be failing to read with sufficient accuracy a particular angle of elevation. An error of detail in the measuring rod technique might be simply forgetting to count the occasional individual measurement. But whether the error is one of principle or one of detail, it is clearly highly improbable that both methods should be substantially in error and yet agree. Suppose, for example, that the wrong theorem of geometry was used in calculating the height of the mountain, and so the calculated figure of 25,807 was very different from the true figure. Suppose also that in counting the number of applications of the foot-rod used, the climbers lost track at some point, and resumed counting at the wrong place, with the result that the obtained figure was wildly wrong as a measurement of the height of the mountain. It would plainly be an extraordinarily unlikely fluke if the climbers produced the same figure of 25,807. We can therefore say it is highly likely that such a fluke has *not* occurred. But the only way it seems to be possible for such a fluke to have not occurred is for both techniques to have measured the height of the mountain (at least more or less) accurately. This, I think, is the reasoning behind why we have much more confidence in measurements that are the result of cross-bracing.

More or less the same reasoning as that given above leads us to the further conclusion that both methods used are likely to have a propensity to be reliable. The reasoning given above establishes that it is likely that in this particular case both methods have led us to a result that is at least *nearly* correct. So, obviously enough, it has shown that one of the methods – say the method of using the theodolite – has in this case led us to a result that is nearly correct. We can now ask the question "Is it merely a fluke that the method of the theodolite has in this case led us to a nearly correct result, or is it rather the case that this method has a propensity to lead us to nearly correct results?" Since it is highly unlikely that it should be merely a fluke that the method has led us to a correct result, we may conclude, by reasoning already defended, that there is a propensity for the theodolite method to lead us to correct results. The same, of course, can also be said for the other method.

Nothing in the argument just given depends upon the details of the particular case considered. We may assert, quite generally, that if *a priori* unlikely agreement has been obtained between two independent methods, then it is highly unlikely that the two methods are substantially in error, since that would mean two *distinct* sources of error happened to lead to *the same* result: itself an occurrence that is *a priori* highly unlikely. And reasoning of a form already defended leads us to the further conclusion that both methods have a propensity to lead us to correct results.

It is worth noting that the two independent methods involved in an AIM inference need not necessarily both be methods of empirical measurement. Sometimes hypotheses are arrived at in science not by measurement or observation, but by plausibility arguments. If some empirical measurements and some plausibility arguments both independently lead to the same conclusion, then we may say that the reliability of both methods has received confirmation.

One example of a plausibility argument and empirical measurements supporting the same conclusion concerns the rates at which objects with different masses fall. An argument was given by Galileo for the conclusion that the rate at which an object falls is independent of its mass. The argument is as follows:

Plausibly, two bricks, both one kilogram in weight, will fall at the same rate. They will fall at the same rate even if they fall placed side by side. But the rate at which they fall will not be affected if they are tied or glued together, instead of merely falling side by side. Two one-kilogram bricks tied or glued together can be regarded as a single object of two kilograms. So a two-kilogram object will fall at the same rate as a one-kilogram object. The argument can be iterated to show that all objects, whatever their masses, will fall at the same rate. Therefore, the rate at which an object falls is independent of its mass.

This (rather *a prioristic*) argument and empirical measurements performed by Galileo both independently supported the same conclusion: that rate of falling is indeed independent of mass. So the AIM inference permits us to say that this increases our confidence in the reliability of *both* the plausibility argument and the techniques of measurement. That the empirical observations supported (to within the limits of accuracy) the claim that rate of falling is independent of mass provides confirmation of the soundness of the plausibility argument. Our increased confidence in the soundness of the plausibility argument in turn increases our confidence that rate of falling is indeed truly independent of mass, and not merely "independent as far as our measurements can tell".

The AIM inference also helps us to understand why scientists should prefer certain theories with low whole numbers. Consider, again, inverse square laws of force. Suppose empirical evidence suggests that the law obeyed by a force is at least close to inverse square. A plausibility argument, relying on the assumption that that which is responsible for the force is conserved as it moves away from its source, independently leads to the conclusion that the law is *exactly* inverse square. The agreement of these two independent methods increases our confidence in the reliability of both methods, in particular in the reliability of the plausibility argument. And so the AIM inference increases our confidence that the value of the exponent is 2 *precisely*.

SYMMETRY

The final success-conducive property of theories we will consider is *symmetry*. Modern physics exhibits a strong preference for theories that are "symmetrical". It should be noted, however, that the term "symmetry" as it used in modern physics only partially resembles the meaning of that term as it is used in ordinary English. Here is a definition of the term "symmetry" as it is used in modern physics:

> A property P of a physical system S exhibits symmetry if and only if P remains unchanged when the physical system S is subjected to some transformation T.

This definition should be taken as *stipulative*, rather than as an explication of some pre-existing notion. In particular, it is not an explication of our ordinary notion of symmetry.

There are two points that need to be noted about this definition. First, symmetry, as defined, is a property *of properties* of systems; it is not, strictly speaking, a property of physical systems themselves. And second, a property is not symmetrical or asymmetrical outright: it is only symmetrical or asymmetrical *with respect to a particular transformation*.

Symmetry, in this sense, includes our ordinary notion of symmetry, but also includes much else. We would ordinarily say that a sphere is symmetrical. And there is also a respect in which a sphere is "symmetrical" on the definition given above: if we rotate a sphere about its centre, in any direction, then the space it occupies will remain exactly the same. The space occupied by a sphere is "symmetrical" with respect to the "transformation" of rotation. In this respect, a sphere differs from an asymmetrical object such as a hand. The space occupied by my hand does *not* remain the same when I rotate it, and so the space it occupies is not symmetrical – in this technical sense of "symmetrical" – with respect to rotation. The everyday notion of symmetry seems to be pretty much synonymous with "symmetry – in the technical sense – with respect to rotation through space".

Although in some cases the technical notion of symmetry coincides with the ordinary notion, in other cases it does not. Consider, for example, a quantity of water at one atmosphere of pressure. This quantity of water has the property of being disposed to boil at 100°C. But this property of being disposed to boil at 100°C at 1 atm is symmetrical with respect to the operation of movement through space; that is, if we move the water from one point in space to another, then so long as the pressure remains the same, the water will retain the property of boiling at 100°C. It is also, presumably, symmetrical with respect to the operation of movement through time. Certainly, so long as the pressure remains constant, if we just let time elapse then, in one hour, or one day, or one year, the water will still have the property of being disposed to boil at 100°C. But it is presumably also the case that if we could move the water around in time as we can in space – if, for example, we could put it in a time machine and send it back to, say, the year 1765, the temperature at which it boils will remain unchanged. The property of boiling at 100°C at one atmosphere of pressure is symmetrical with respect to the operations of "translation" through space and time.

Another type of symmetry arises from the operation of *replacement*. For example, if one electron in a physical system is replaced with another, then the physical properties of the system remain unchanged: the properties of the system are symmetrical with respect to the operation of replacement of one electron by another.

Plainly, the symmetry of *a system* is something that can come in degrees. This follows from the definition of symmetry: a property P of a system is symmetrical with respect to some transformation T if and only if T leaves P unaltered. So a system can be said to be highly symmetrical if it has *many* properties that exhibit symmetry, and a property of a system can be said to be highly symmetrical if there are many transformations that the system of which it is a property can undergo while leaving the property unaltered.

Theories that attribute to the physical world a high degree of symmetry are much sought after in contemporary theoretical physics. It will be argued that this preference is naturally explainable within the framework developed here.

There are also two more familiar concepts with which the notion of symmetry is clearly closely related; these are induction and conservation. Consider an inductive generalization such as "All electrons have negative charge." This is usually taken to mean at least: "Any electron, at any point of space and time, has negative charge." If it is taken to be an explanatory law, it is also usually taken to also entail: "If any electron were, for example, to occupy a different point of space and time from that which it actually occupies, it would still have negative charge." It is usually, I think, also taken to entail that if a new electron were to come in to being, it too would have negative charge. All these claims are consequences of the idea that electrons have a propensity to have negative charge. But all these assertions make claims about the existence of various *symmetries*. They assert that the property of being an electron is associated with the property of having negative charge even if the point in space and time at which the property of being an electron is instantiated is altered, or if an electron is moved from one place in space and time to another, or if new instantiations of the property of being an electron arise. The fact that the property of being an electron is associated with the property of having negative charge is something that remains unaltered after these changes, and so the fact that these two properties go together is something that exhibits a high degree of symmetry.

All these symmetries are, as we have noted, logical consequences of electrons having a *propensity* to have negative charge. But the existence of such a propensity is warranted by the account advocated here: if we have observed many electrons that all have negative charge, then we are justified in concluding there is a propensity for electrons to have negative charge. The existence of propensities can explain the existence of a range of symmetries. So we can, in this way, explain why scientists prefer *some* symmetrical theories.

The notion of symmetry also bears a close relation to the notion of the *conservation* of physical magnitudes. Indeed, the considerations that led us to say that conservation laws are especially preferable also apply to laws that postulate symmetries. To say that a symmetry exists is to say that a particular property remains unchanged under a certain transformation. But to say that that property remains *unchanged* is to say that the size of the change the property undergoes is zero. But the reader will recall that the feature of conservation laws that was responsible for their "privileged" status is that they postulated zero change. And so the arguments that led us to say that conservation laws are especially privileged also apply to laws postulating symmetries.

CONCLUDING REMARKS

In this chapter a number of additional success-conducive properties of theories have been described. The first of these was intra-DEC independence. It should be noted, however, that the notion of intra-DEC independence was *not defined*. Some sufficient conditions for one theory having a higher degree of intra-DEC independence than another were given, but no claim was made that these conditions were also necessary.

It was also argued that there was one type of theory that possessed the highest degree of intra-DEC independence of all: the conservation laws.

It also sometimes occurs in science that we infer from the fact some measured quantity is close to a low whole number that the true value of the quantity is precisely equal to that low whole number. It was argued that there are certain conditions under which we make this inference, and that making the inference under those conditions can be justified within the general framework advocated here.

Finally, it was argued that laws postulating symmetries are preferable for the same reason as laws that say some magnitude is conserved are especially preferable: both assert that zero change occurs, whether over time or as a result of some kind of transformation. Both conservation laws and those asserting the existence of a certain kind of symmetry have an especially high degree of "this-couldn't-be-due-to-chance-ness".

CHAPTER 6

Newton's laws of motion and law of gravitation

The overall aim of this book is to explain certain forms of scientific success. One set of theories that exhibit all three of the forms of success with which we are here concerned is Isaac Newton's three laws of motion and law of gravitation. Newton's laws have been used in the derivation of numerous novel predictions, including the derivation of the prediction of the existence of the outer planets, their positions at a particular time, their orbit paths, their momenta and their departure from perfect sphericity. So Newton's laws clearly exhibit the first phenomenon. As we have already observed, they have also led us to a knowledge of some parts of reality, such as the planet Neptune, that were not accessible at the time they were first postulated, but whose existence is now beyond serious dispute. So they exemplify the second of our phenomena. Finally, at least some of Newton's laws are, arguably, close to *a priori*, and so exhibit the third phenomenon.[1] Since Newton's laws exemplify all three of the types of success with which we are here concerned, it is highly desirable that the account advocated here be able to explain the successes of those laws.

NEWTON'S THREE LAWS OF MOTION

Newton's three laws of motion (Newton 1960: 13)[2] are:

1. Every body continues in its state of rest, or of uniform motion in a straight line, unless it is compelled to change that state by forces impressed upon it.
2. The change of motion is proportional to the motive force impressed and is made in the direction of the straight line in which that force is impressed.
3. To every action there is always opposed an equal reaction, or the mutual actions of two bodies upon each other are always equal and directed to contrary parts.

Here it will be argued that the acceptance of these laws can be explained on the views advocated here. Let us start by considering Newton's first law. One thing about this law is that we can be sure that no one, including Newton, has ever observed any positive instances of it. All the objects we see in the world around us are acted upon by some force or other. So it is initially not clear what empirical grounds Newton could have had for asserting his first law. However, in the "Scholium" attached to his introductory chapter on the laws of motion, Newton states that the principles he has laid down "have been received by mathematicians, and are confirmed by abundance of experiments". But the Scholium also contains qualifying statements such as "unless so far as these motions are a little retarded by the resistance of the air" (Newton 1960: 21), "as far as I can perceive" (Newton 1960: 25), and so on. That Newton should see the acceptance of the laws by *mathematicians* (as opposed to experimentalists) as evidence of their correctness perhaps suggests that Newton sees the laws as having some kind of *a priori* status. However, if they are *a priori* truths, they are *a priori* truths that have also received confirmation by observation and experiment. Moreover, as Newton acknowledges, the observations do not exactly confirm the laws. The difference between what the laws predict and the observed behaviour of bodies is to be explained by factors such as "the resistance of the air", the lack of perfect elasticity of bodies and the inevitable imprecision of our measurements.

It is, of course, possible for a sceptic to give a different interpretation of the lack of perfect match between what the laws predict and our actual observations. Perhaps the lack of perfect agreement between our observations and the predictions of the first law are not merely due to, for example, friction or gravity. Perhaps it is *false* that a body not acted on by any force remains at rest or moving in a straight line at constant velocity. Perhaps such a body is not at rest, but jiggling slightly or imperceptibly; or perhaps its path is not an exact straight line but a curve that deviates imperceptibly from a straight line. These possibilities were certainly not ruled out by any observations available to Newton, and it is plain that *some* hypotheses of this sort could never be ruled out by any observation we are capable of making. Yet Newton apparently accepted that bodies do remain *at rest*, or moving in a *straight* line.

This is, of course, explicable on the account offered here. Observations available to Newton indicated that bodies not acted on by any force came at least very close to behaving as described by the first law. The observations available to Newton were *compatible* with the first law, although they were also compatible with slight deviations from it. But there were also other lines of reasoning available that led *independently* to the first law. Suppose a body not acted on by any force was imperceptibly jiggling. Then the momentum of the body, or what Newton called "quantity of motion", would not be remaining constant, but would be continually increasing and decreasing. This would, of course, be incompatible with the idea that "quantity of motion", or momentum, should be conserved. But as we noted in Chapter 5 (pages 102–4), the hypothesis that a quantity is conserved is more likely to lead to correct predictions than any other.

Much the same considerations apply to the second half of the first law: that a body in motion, not acted on by any external force, remains moving at a constant velocity in a straight line. Suppose a body A is already in motion. There are three ways it could depart from moving at a constant velocity in a straight line. These are:

1. its velocity could change;
2. its direction of motion could change; or
3. both (1) and (2).

Plainly, if its velocity changes, then neither its momentum nor its kinetic energy is conserved. But if its velocity remains constant, then momentum and kinetic energy are conserved. And so we are again led to the conclusion that the hypothesis that the velocity does not change will be more likely to be lead to correct predictions than any alternative hypothesis.

It is easy to see that the same type of argument also leads us to prefer the hypothesis that the body will continue to move in a straight line. Plainly, it is possible for a body to move in a curve at constant velocity. But if it is moving in a curve, although its "total" velocity may remain constant, the components of its velocity in at least some of the x, y and z axes will change. But if it remains moving in a straight line, the components of its velocity in all those axes will remain constant. So, again, the above arguments lead us to conclude that the hypothesis that is most likely to lead to correct predictions is the hypothesis that the body continues to move in a straight line. Consequently, we may assert that the arguments developed in earlier chapters show that, if the obtained observations are in close agreement with Newton's first law, then it is rational to assert that Newton's first law is more likely to lead to correct predictions than any alternative hypothesis also in close agreement with the obtained observations. Moreover, by the argument of Chapter 5 (pages 105–7), we can also account for its close-to-*a priori* character.

Newton's second law states: "The change of motion is proportional to the motive force impressed and is made in the direction of the straight line in which that force is impressed." We should note here that by "motion", Newton means "momentum" rather than "velocity". Perhaps the first thing that strikes us about this law is its extreme intuitive plausibility. We are inclined to say that it, or something roughly like it, is "just common sense". *Of course*, we are inclined to say, the harder we push something, the greater the change in motion we will produce in it, and *of course* pushing it in a particular direction will make it move in that direction, rather than some other. Newton's own brief commentary on the law also brings out this intuitive plausibility. He writes, "If any force generates a motion, a double force will generate double the motion, a triple force triple the motion, whether that force be impressed altogether and at once or gradually and successively." (Newton 1960: 19). Here we can see Newton offering us what might with some plausibility be seen as an *a priori* argument for the first part of the law, that is the part of the law which says "the change is proportional to the motive force impressed". Recall that what Newton means by "motion" is what we now mean by "momentum". Suppose that on one occasion a particular force F of magnitude m acts on a body and increases its momentum by p. It is surely very plausible to say that if F once again acts on the body on a later occasion, it will again increase its momentum by p. That is, if double the original force F acts on the body "successively", that is if F acts on one occasion and then F again acts on another, the total increase in momentum will surely just be $p + p = 2p$. But now suppose that the two units of F act not successively but simultaneously. It is *a priori* very plausible to think that whether they act one shortly after the other or both at the same time, their total effect will be the same. So, if two units of force F act simultaneously, the total increase in velocity will again be $2p$. The argument could be repeated for three units of F, four units and so on. And so we plausibly have a more or less *a priori* argument for the plausibility of the first half of the second law, that is for the part of the second law which asserts that the change in motion is proportional to the motive force impressed. It should be noted that nothing depends on the plausibility of this argument being strictly *a priori*: the point is that it is an argument which surely has some degree of plausibility leading us to the conclusion that the change in motion is proportional to the motive force impressed. That is it is an argument, independent of precise empirical observations, leading us to the conclusion that the first half of the second law is correct. It also seems to be an argument sketched in Newton's own comments on the second law. But, of course, the more precise observations available to Newton were consistent with the claim that the first half of the second law was correct. Therefore, the AIM inference leads us to the conclusion that the claim that "change of motion is proportional to motive force impressed" is more likely to lead to correct predictions than any alternative claim. And since the argument used by Newton is plausibly close-to-*a priori*, we can account for the close-to-*a priori* character of the law.

What is the relation between the *direction* of the "motive force impressed" on an object and the *direction* in which the object moves? Actually obtained results are consistent with the relation being that of identity, but they are also consistent with very slight differences between the two directions. Saying that there is *no* difference between the two directions plainly leads to a more parsimonious DEC than any alternative hypothesis also compatible with actually obtained observations. So, by the argument of Chapter 5 (page 98), we are able to account for the second law.

Newton's third law states that "To every action there is always opposed an equal reaction, or the mutual actions of two bodies upon each other are always equal and directed to contrary parts." Unlike, for example, the second law, which seems to be a precise statement of something we are all inclined to regard as "common-sense", the third law does not at first appear to be intuitively compelling. However, he goes on to write:

> If a body impinge upon another, and by its force change the motion of the other, that body also (because of the equality of the mutual pressure) will undergo an equal change, in its own motion, towards the contrary part. The changes made by these actions are equal, not in the velocities but in the motions [momenta] of bodies.　　　　　　(Newton 1960: 19)

So Newton is here saying that when one body "impinges" on another, and "by its force changes the motion of the other", the fact that the actions that each body produces on the other are equal, but in opposite directions, will mean that the change in momentum that each body produces in the other will be equal, but in opposite directions. Newton presents this as a consequence of the third law. But the fact that this is a consequence of the third law can also be seen as an argument for it. Consider two bodies "impinging" on, or interacting with, each other. It is a consequence of the third law that any change in the momentum p of the first body will be accompanied by an equal and opposite change $-p$ in the momentum of the second body. That is, the *total* momentum of the two bodies together will remain unchanged after the interaction. Therefore it is a consequence of the third law that in any interaction between two bodies, momentum will be conserved. Moreover, if the action of the first body on the second was not accompanied by an equal and opposite reaction, the changes in momentum would not be equal and opposite, and so momentum would not be conserved. Consequently, if and only if the third law is true will momentum be conserved whenever two bodies interact. But we have already given a number of reasons why conservation theories are more likely to lead to correct predictions than alternatives equally compatible with the obtained observations.

In summary, therefore, it can be seen that Newton's three laws of motion are supported by the argument from the preferability of theories with high intra-DEC independence and the AIM inference. Although an indefinitely large number of other laws would be consistent with the empirical data available to Newton, and would in fact be consistent with any body of empirical data of a finite degree of accuracy, these two arguments show that Newton's laws are the ones that are most likely to be lead to correct predictions.

NEWTON'S "RULES OF REASONING IN NATURAL PHILOSOPHY"

Newton's *Principia* is divided into three books. He lays down his three laws of motion at the beginning of the first book. The whole of the first book and the second book are, in Newton's words, concerned with "principles not philosophical but mathematical". The aim of the first two books is, that is, to derive mathematical consequences of the three laws. It is not until Book Three, the "System of the World", that Newton uses the principles derived in the first two books to give a description of the how the world is. More specifically, in Book Three Newton develops his account of the Solar System. Book Three also contains his argument for universal gravitation. But at the beginning of Book Three he lays down four "rules of reasoning in philosophy". We would perhaps be more inclined to call them "rules of inductive inference" or "rules for inferring the best explanation". In this section it will be argued that these four "rules of reasoning" can be seen as strategies for increasing the independence of theory from data".

Newton's four rules of reasoning (Newton 1960: 398–400) are as follows:

1. We are to admit no more causes of natural things than such as are both true and sufficient to explain their appearances.
2. Therefore to the same natural effects we must, as far as possible, assign the same causes.
3. The qualities of bodies, which admit neither intensification nor remission of degrees, and which are found to belong to all bodies within reach of our experiments, are to be esteemed the universal qualities of all bodies whatsoever.
4. In experimental philosophy we are to look upon propositions inferred by general induction from phenomena as accurate or very nearly true, notwithstanding any contrary hypothesis that may be imagined, till such time as other phenomena occur, by which time they may either be made more accurate or liable to exceptions.

On the face of it, Newton's first rule would appear to be a principle akin to Ockham's Razor. The gist of the first rule can perhaps be stated as, "We are entitled to assert the existence of some cause if it is necessary to explain some observations or appearances, but if some hypothetical cause is not needed to explain some observations or appearances, we are not entitled to assert its existence." Quite plausibly, it is also implied by the principle that if we are to admit no more causes than are sufficient, we are to *accept* that explanation that minimizes the number of causes postulated. If this is an accurate account of the first rule, then it is plain that it can be seen as a way of maximizing the independence of our theory from the data. We maximize the independence of theory from data by minimizing the ratio of dependent explanatory components of theory to components of data to be explained. Plainly, then, "postulating no more causes of natural things than such as are … sufficient to explain their appearance" will have the effect of keeping the number of DECs at a minimum, and thereby keeping the independence of theory from data at a maximum.

It should be noted that there is one feature of Newton's first rule that seems a little odd and calls for some comment. Newton says that we are to admit no more causes than such as are both *true* and sufficient to explain appearances. The requirement that we are only to admit *true* causes seems very strange, since the very point of the rules of reasoning seems to be to provide us with rules of inference that will lead us to true conclusions. If we already know which causes are true, why do we need the "rules of reasoning"? I suggest that we can make sense of what Newton is saying here if we assume that by "true" Newton here means what we would more usually mean by "unrefuted" or "unfalsified" or "not yet shown to be false". Or Newton might be taking "true" to mean "empirically adequate" or "empirically adequate as far as we know". On this view, what the first rule is saying is that we should admit those causes of appearances that have not yet been shown to be false and which are the minimum necessary in order to explain the appearances. Interpreted in this way, Newton's rule is plainly in agreement with the view advocated in Chapter 4 (pages 67–8), where it is argued that the best hypothesis is the one, from among those that are not refuted by the data, that maximizes the degree of independence from the data.

Newton's second rule is: "Therefore to the same natural effects we must, as far as possible, assign the same causes." It is plain that this rule also can be seen as a strategy for increasing the independence of theory from data. Suppose we have a number of instances of the same natural effect. It is plain that a theory that explains all these instances by just the same type of cause will have fewer DECs than will a theory that explains the instances with a number of different types of cause.

It is worth noting that Newton's statement of the law contains the phrase "as far as possible". Presumably this qualification is included since we might obtain some empirical results that show that in some cases similar effects are actually due to different causes. So, plausibly, Newton is asserting that we ought to assume that similar effects have similar causes until that hypothesis is refuted by empirical observations. So, on the interpretation offered here, both the first rule and the second rule say we ought to accept certain types of hypotheses provisionally, or unless refuted by other or subsequent observations. Newton's use of the word "therefore" at the beginning of his statement of the second rule suggests that he sees it as a consequence of his first rule. This provides further evidence that when, in his statement of the first rule, Newton says we are only to admit "*true*" causes, what he means is that we are only to admit causes which have as yet not been refuted.

Newton's third rule is: "The qualities of bodies, which admit neither intensification nor remission of degrees, and which are found to belong to all bodies within the reach of our experiments, are to be esteemed the universal qualities of all bodies whatsoever." This is plainly a principle of induction, but it is a carefully qualified principle of induction. Newton is plainly *not* saying that if all observed A have been B, then we may conclude that *all* A are B. He is rather saying that we are only entitled to draw this conclusion in those cases in which the property A is one that "admits of neither intensification nor remission of degrees". Presumably, the reason for this qualification is as follows. Suppose that all the A we had observed so far had been B, but the property B did admit of "intensification or remission of degrees". Then, it seems reasonable to say, there could be some A somewhere that lacked B altogether. Hence, in such a case, we would not be entitled to draw the conclusion that *all* A were B. For example, suppose all the bodies we had observed so far had a length of at least one-hundredth of one inch, although there was also a great variation in the length of different bodies. (The length of bodies did admit of intensification and remission of degrees.) Then, on Newton's third rule, we would not be entitled to draw the conclusion that all bodies had a length of at least one-hundredth of an inch. But, on the other hand, the property of having some length or other, or the property of having *some* length rather than no length at all, is a property that does not admit of intensification or remission of degrees, and so, on Newton's third rule, we would be entitled to draw the conclusion that all bodies have *some* length.

We have already noted that the drawing of conclusions by induction, and, more particularly, the use of the straight rule, is an important means for increasing the independence of theory from data. However, it should also be recalled that on the view defended here, we are not entitled to accept inductive generalizations as true, but only as likely to lead to correct predictions. On the face of it, Newton appears to be making the stronger claim that we are indeed justified in accepting the results as *true*. But we have also noted that there is some evidence that Newton understands the term "true" to mean something like "empirically adequate". If this is correct, Newton's position would seem to be the same as the one advocated here.

As we have observed, Newton does not advocate the straight rule *simpliciter*: he advocates a version of it restricted to properties that admit of neither intensification nor remission of degrees. Can this restriction be explained within the general framework advocated here? The answer is "Yes it can." The basic inference of science, on the view advocated here, is of the following form:

Premise 1: It is unlikely that result E should have been obtained by chance.
Therefore: It is likely that it was not just due to chance that E was obtained.
Therefore: There is a tendency or propensity for E to be obtained.

It will be argued that premise 1 in the above argument is unlikely to be correct if it concerns a property that *does* admit of intensification or remission of degrees. Consider, for example, a case we have already mentioned: the property of having some length greater than some particular length, such as one-hundredth of an inch. Suppose that objects we have looked at have come in all sorts of sizes. Plainly, if we have only observed a finite number of objects, then, necessarily, one of the objects we have looked at will be the smallest. Suppose the length of the smallest object is n inches. Then, since the objects we have looked at have come in all sorts of sizes, it may very well just be due to chance that the smallest object we have looked at happens to be n inches long. Consequently, we would not be justified in drawing the conclusion that it was not due to chance, and so the argument for the existence of a propensity for objects to be at least n inches long could not get started.

It is also worth noting that if, counterfactually, the property of length was one that did not admit of intensification or remission of degrees, the argument advocated here would go through. Suppose all the objects we had looked at had proved to be exactly m inches long, no more and no less. If we had looked at a large number of objects, it would be reasonable to conclude that it was unlikely that it was just due to chance that they should all be exactly m inches long, and so the argument to the conclusion that there was a propensity for objects to be exactly m inches long would go through.

In summary, Newton's third rule of reasoning is a consequence of the framework developed here.

It is worth briefly noting the reasons Newton himself gives in support of this rule. He says, "We are certainly not to relinquish the evidence of experiments for the sake of dreams and vain fictions of our own devising; nor are we to recede from the analogy of nature of Nature, which is wont to be simple, and always consonant to itself" (Newton 1960: 398–9). The first part of what Newton says here seems to be not so much a reply to the *sceptic* as a reply to those who are inclined to construct intellectual castles in the air. He seems to be saying that if theories based on experimental evidence conflict with mere speculations, we are to accept the theories based on evidence and reject the mere speculations. The second part of what Newton says appears to be an attempt to "justify induction" by appeal to a principle of simplicity and to the uniformity of nature. I concede that there is nothing in what Newton says that can be construed as an attempt to justify the third rule in the way that has been advocated here. However, this need not count too heavily against the view defended here. Although it is a part of the view defended here that scientists have some kind of knowledge of the arguments justifying induction and a preference for highly independent theories, this knowledge need not be explicit knowledge. It might be merely implicit, or practical, or "inchoate" knowledge. And, of course, it can be very difficult to translate implicit or practical knowledge into theoretical knowledge, as is shown by the difficulty of constructing an explicit theory of the knowledge we all use when we recognize a sentence as grammatical.

Newton's fourth rule of reasoning is: "In experimental philosophy we are to look upon propositions inferred by general induction from phenomena as accurate or very nearly true, notwithstanding any contrary hypotheses that may be imagined, till such time as other phenomena occur, by which they may either be made either more accurate or liable to exceptions." To modern ears, this sounds like a statement of a rather confident epistemological optimism. And the point of view expressed in this rule does seem to be somewhat more optimistic than the one defended here. For example, Newton's *first* rule says, very roughly, that we are to prefer the simplest theory. When this is conjoined with the fourth rule, we see Newton is recommending that we accept the simplest hypothesis as true or very nearly true. This is rather more optimistic than the view advocated here, which merely says that we should accept that the simplest (or the most independent) theory is more likely than other theories to continue to lead us true beliefs *about the data*. But two points need to be noted. First, it is clear that Newton only intends this acceptance to be *provisional*: the hypotheses inferred by induction are only to be accepted as (nearly) true until further observations show they need to be rejected or modified. And second, as was the case with his third rule, Newton does not seem to see the main function of this rule as furnishing a reply to the *sceptic*; rather he seems to see it as providing a reply to those who would have our beliefs determined by fanciful speculation rather than observation and experiment. He says, "This rule we must follow, that the argument of induction may not be evaded by hypotheses." When we note the function that Newton wishes this rule to play, the point of view he expresses is perhaps not too violently in conflict with the one defended here. It is also appropriate to note that, in our discussion of Newton's first and second rules, we observed that there was some evidence that Newton perhaps understood the term "true" to mean something like "empirically adequate". If this is correct, then the point of view Newton adopts in his fourth rule is actually in close agreement with that advocated here.

In summary, Newton's first three rules of reasoning can all be seen as strategies for increasing the independence of theory from data. Newton's restriction of the third rule to qualities which "admit neither intensification nor remission of degrees" is easily explainable within the view advocated here. In the fourth rule, Newton perhaps expresses a degree of epistemological optimism that goes beyond that accepted here, but the function of the fourth rule seems to be as much to restrain the fanciful speculator as it is to reply to the sceptic. Moreover, if Newton has some sort of empiricist rather than full-blown realist conception of truth – and we saw in our earlier discussion that perhaps he does – the fourth rule is actually in close agreement with our position.

NEWTON'S ARGUMENT FOR UNIVERSAL GRAVITATION

Newton begins his argument for universal gravitation by listing six "phenomena", the correctness of which he takes to have been established by astronomical observations. He then uses these six phenomena, together with his "rules of reasoning in natural philosophy" to derive his conclusion that all bodies have a "power of gravity", that this power is proportional to the masses of the bodies and that it varies inversely with the square of the distance from the body.

The six phenomena that constitute the starting point of Newton's argument (Newton 1960: 401–5) are as follows:

1. That the moons of Jupiter obey the analogues of Kepler's second and third laws of motion.
2. That the moons of Saturn also obey the analogues of Kepler's second and third laws.
3. That the five primary planets (that is Mercury, Venus, Mars, Jupiter and Saturn) orbit the Sun.
4. That the five primary planets obey Kepler's third law, as does the Earth (if it is interpreted as orbiting the Sun) and the Sun (if it is interpreted as orbiting the Earth).
5. That the five primary planets obey Kepler's second law if taken to orbit the Sun but do not do so if taken to orbit the Earth.
6. That the Moon obeys Kepler's second law.

As the six phenomena make repeated reference to Kepler's three laws of planetary motion, it is useful to remind ourselves of them:

> That the planets move around the Sun in an ellipse with the Sun at   (K1)
> one focus.

That a line drawn from any planet to the Sun will sweep out equal   (K2)
areas in equal times as the planet orbits the Sun.

That for any two planets, the ratio of their orbit times is equal to the   (K3)
½ power of their mean distance from the Sun.

As we stated above, Newton himself accepts the six phenomena. But it is worth briefly considering *why* he accepts them. We will not consider the case for each of the six phenomena individually; the points made in connection with the first of the phenomena apply also to the other five.

Newton says the orbits of the moons of Jupiter "differ but insensibly from circles concentric to Jupiter" (Newton 1960: 401). I will assume this to mean that, given the accuracy of our observations, the orbits of the moons of Jupiter could be circles, but could also be other shapes slightly different from circles. Plainly, Newton takes this as showing the orbits are indeed *true* circles, and therefore ellipses. But the observations are also compatible with the orbits differing imperceptibly from true circles. We can explain Newton's willingness to accept that they are true circles as being due to an implicit use of the argument to the maximally simple DEC, that is the DEC corresponding to a conservation law. To say that the orbits of the moons of Jupiter are circles is to say that over any period of time the distance from Jupiter of any moon does not change. Any alternative to saying the orbits are circular will say that over time there is some non-zero change in the distances. Therefore, by considerations given on page 102, asserting that the orbits are circular gives us a description with a higher degree of intra-DEC independence than any other, and hence a DEC that is more likely to lead to correct predictions.

Newton says that a range of astronomical observations confirm that the orbit times of the moons of Jupiter "are as the ½ power of their distances from its centre", that is that they obey the analogue of Kepler's third law (Newton 1960: 401). It is worth looking at the astronomical evidence Newton actually gives us in support of this claim. We will restrict ourselves to the innermost of the moons of Jupiter. The distance of the innermost moon had been calculated by a number of different scientists in different ways. The results are given below:

- *Borelli* found the distance to be 5⅔ the radius of Jupiter.
- *Townly*, using the micrometer, found it to be 5.52 times the radius of Jupiter.
- *Cassini*, using the telescope, found the distance to be 5 times the radius of Jupiter.
- *Cassini*, using the eclipse of the satellites, found it to be 5⅔ the radius of Jupiter.

Newton then goes on to note that, given the orbit time of the innermost moon, Kepler's third law enables us to predict that the distance of the innermost moon from Jupiter should be 5.667 times the radius of Jupiter. That is, *if* the innermost moon obeys Kepler's third law, its distance should be 5.667 radii. Newton takes the close agreement between the predicted value and the values obtained by observation to show that the moon does indeed obey Kepler's third law.

Perhaps our first response is that the values cited do not appear at all to show the moon to obey Kepler's third law. Two of the values cited come very close to that predicted by Kepler's third law, but the two others differ noticeably from it, one of them by quite a substantial margin. The average of the four values obtained by observation is 5.4635 – somewhat different from the predicted value of 5.667. So why is it that Newton does not draw the conclusion that perhaps the innermost moon of Jupiter does not obey Kepler's third law, but perhaps some slightly different law? One possible answer is suggested by the argument from the AIM inference. It was known from phenomenon 4 that the five primary planets obey Kepler's third law, as does the Earth, if it is taken to orbit the Sun, and the Sun, if it is taken to orbit the Earth. This suggests a more general hypothesis: that orbiting bodies generally obey Kepler's third law. And this in turn leads us to the prediction that the moons of Jupiter will also obey this law. The actually obtained observations come close to those values predicted by Kepler's third law, but are not precisely the same as those predicted values. And so we have a situation in which the AIM inference applies, leading us to the conclusion that the hypothesis that the moons of Jupiter obey Kepler's third law is more likely to lead to correct predictions than alternative hypotheses.

Newton's acceptance of the other five phenomena can all be explained on the views advocated here. All of them, apart from phenomenon 3, can be seen as the result of the application of the argument from the AIM inference, and the argument to the simple DEC, to the empirical data available to Newton. We need not go through all of them in detail here, as our treatment would be essentially the same as that for phenomenon 1. Newton's argument for phenomenon 3 is, however, a little different. He uses the phases exhibited by the planets, and the way they change their brightness as they change their position, to argue both for the conclusion that they do not have their own light but "borrow" or reflect it from the Sun and that they also orbit the Sun (Newton 1960: 404). Newton here seems to be applying his own second "rule of reasoning": objects here on Earth, such as balls or oranges, would change their phases and brightness in the same way if they were moving around, for example, a candle in a darkened room, and so we may conclude that the similar appearances exhibited by the planets are due to a similar cause. But we have already noted that Newton's second rule can be seen as a strategy for increasing the independence of theory from data. So, in summary, we may conclude that Newton's acceptance of the six phenomena can be explained on the view advocated here.

Having made a case for the reality of the six phenomena, Newton then develops his argument for universal gravitation. He does this by deriving a series of "Propositions" from the phenomena. The first three of these Propositions are:

> That the forces by which the circumjovial planets [i.e. the moons (I) of Jupiter] are continually drawn off from rectilinear motions, and retained in their proper orbits, tend to Jupiter's centre, and are reciprocally as the squares of the distances of the places of those planets from that centre.

> That the forces by which the primary planets are continually drawn (II)
> off from rectilinear motions, and retained in their proper orbits,
> tend to the Sun, and are reciprocally as the squares of the distances
> of those planets from the Sun's centre.
>
> That the force by which the Moon is retained in its orbit tends to the (III)
> Earth, and is reciprocally as the square of the distance of its place
> from the Earth's centre.

Propositions I and II appear to follow deductively from some of the phenomena he has cited, and from results established earlier in the *Principia*. For example, proposition I follows from phenomenon 1, the already established "proposition 2" of Book I (which asserts that any object that obeys Kepler's area rule around a point p must be acted on by a force always directed towards p), and corollary 6 to proposition 4 of Book 1 (which asserts that if some body or bodies obey Kepler's third law as they move around p, the force directed towards p will vary inversely with the square of the distance from p). These propositions Newton claims to have been deduced by mathematical and geometrical techniques from his three laws of motion. So we may represent Newton as presenting the following, deductively valid argument for the first part of his proposition I:

> Premise 1   For any *x*, if *x* is a body that obeys Kepler's area rule about point
>   p, then *x* is acted on by a force directed towards p.
> Premise 2   The moons of Jupiter obey Kepler's area rule about Jupiter.
> _____
> Conclusion: The moons of Jupiter are acted on by a force directed towards
>   Jupiter.

It is easy to see that it is also possible to construct a deductively valid argument, from results already established, for the second part of Newton's proposition I. Of course, the use of deductively valid arguments is perfectly compatible with the view defended here. It is only a part of the view defended here that those *ampliative* inferences that lead to subsequent novel success should increase the independence of theory from data, or be defensible by the AIM argument. Moreover, we have seen that the various inferential steps that Newton uses in arriving at his proposition I are of those types, or else are deductively valid. Therefore, the steps by which Newton arrives at proposition I are in accord with the view advocated here.

Newton's argument for proposition II is also a purely deductive inference from results already established.

However, we find Newton making an ampliative inference when arriving at the second part of proposition III: that "the force by which the Moon is retained in its orbit is reciprocally as the square of the distance of its place from the Earth's centre" (Newton 1960: 407). Earlier, Newton had established that if the one body is orbiting another, and the forces of the two bodies obey an inverse square rule, then the position of the apogee of the orbiting body should remain the same from one orbit to another. But observations show that the apogee of the Moon does not remain in a constant position. Newton says that one possible way of explaining the motion of the apogee of the Moon is to say that the force acting on the Moon, and directed towards the Earth, does not diminish with the inverse of the *square* of the distance from the Earth, but rather diminishes with the inverse of the $2^{4}/_{243}$rd power of the distance from the Earth. But another explanation is that the force acting on the Moon does obey the inverse square rule, and the motion of the apogee of the Moon is actually due to the additional gravitational influence of the Sun. And it is this latter explanation that Newton says ought to be accepted.

Newton's preference for this second explanation can be explained if we see him as trying to maximize the independence of theory from data. Newton has already established that the centripetal forces around both the Sun and Jupiter obey the inverse square law. If it is said that the force acting on the Moon directed towards the Earth's centre also obeys this rule, then it will be possible to explain the motions of bodies orbiting the Sun, Jupiter and the Earth with the same law: specifically, the inverse square law. But if it is asserted that the centripetal force around the Earth does not obey the inverse square law, but some slightly different law, then the motions of objects orbiting those three bodies would not be explained by one law, but two. Consequently, the way to minimize the ratio of DECs to components of data explained, and therefore to maximize the independence of theory from data, is to explain the motions using only the inverse square law, and to attribute the motion of the apogee of the Moon to the additional influence of the Sun. Hence, Newton's preference for this hypothesis is explainable if we say he is aiming to maximize the independence of theory from data.

Proposition IV states: "That the Moon gravitates towards the Earth, and by the force of gravity is continually drawn off from a rectilinear motion and retained in its orbit." In his argument for the truth of this proposition, Newton makes it clear that what he is primarily concerned to establish is that "the force by which the Moon is retained in its orbit is that very same force which we commonly call gravity", or that the force we are acquainted with in everyday life and which causes objects to fall to the ground is the very same force as that which is keeping the Moon in its orbit.

Newton's argument for the identity of these forces proceeds in several steps. First, he reminds us that in proposition III it was established that "the force by which the Moon was retained in its orbit tends to the Earth". That is, there is some force acting on the Moon and directed towards the Earth. He then reasons that if the Moon were "deprived of all motion", then this force acting on the Moon and directed towards the Earth would cause the Moon to descend towards the Earth. Newton is able to calculate how quickly the force would cause the Moon to descend towards the Earth, given that the force is strong enough to "retain the Moon in its orbit" and prevent it from flying off into space. But now, in proposition III, Newton also argued that it is rational to assert that the force directed towards the Earth, retaining the Moon in its orbit, obeys an inverse square law. Given that he knows how quickly the force would cause something as far as away as the Moon to descend, and given that the force obeys an inverse square law, Newton is then able to calculate how quickly the force would cause an object at the surface of the Earth to descend. He observes that the force would cause objects at the surface of the Earth to descend at exactly the *same* rate that we see familiar objects around us falling due to the force of gravity. From this Newton concludes that the force keeping the Moon in its orbit is actually one and the same force as gravity. He writes:

> And therefore the force by which the Moon is retained in its orbit becomes, at the very surface of the Earth, equal to the very same force we commonly call gravity. And therefore (by rules 1 and 2) the force by which the Moon is retained in its orbit is that very same force which we commonly call gravity. (Newton 1960: 408)

It is clear that the independence of theory from data is maximized by saying that the force retaining the Moon in its orbit is one and the same force as that responsible for the descent of objects such as sticks and stones here at the surface of the Earth. This can be seen if we contrast it with some other possible ways of explaining the data. It could, for example, be that the two forces are numerically distinct: one force operates here at the surface of the Earth, at some point between the Earth and the Moon this first force ceases to operate, but another force comes into operation in the region of the Moon. But postulating two forces rather than just one would plainly increase the ratio of DECs to components of data explained and so reduce the independence of theory from data.

It should also be noted that postulating two *distinct* forces would give rise to what would seem to be a remarkably improbable fluke. Newton had argued that the force retaining the Moon in its orbit obeyed the inverse square law, or, at least, it obeyed that law in the vicinity of the orbit of the Moon. But Newton has established that if that force, obeying an inverse square law, *were* to extend down to the surface of the Earth, it would there have exactly the same strength that gravity has at that location. Plainly, if the two forces really were distinct, the identity of their strengths would be a highly improbable fluke. It is the type of thing of which we would say "this could not be due to chance", and so would conclude it was highly likely that it was not due to chance. But if we say that the two forces are in fact one and the same, we explain why their strengths, at the surface of the Earth, should be identical. Postulating the identity of the two forces turns what would otherwise be a highly improbable fluke in to precisely what is expected.

One important implication of proposition IV is that if the force retaining the Moon in its orbit obeys the inverse square law, and that force is one and the same as gravity, then gravity must obey the inverse square law.

Proposition V asserts: "That the cirumjovial Planets gravitate towards Jupiter; the circumsaturnal towards Saturn; the circumsolar towards the Sun; and by the forces of their gravity are drawn off from rectilinear motions, and retained in curvilinear orbit." Newton's argument for this proposition is that it follows from Proposition IV together with rule 2. He says that "The revolutions of the circumjovial Planets about Jupiter, of the circumsaturnal about Saturn, and of Mercury and Venus and the other circumsolar Planets about the Sun are appearances of the same sort with the revolution of the Moon about the Earth; and therefore, by rule 2, must be owing to the same sort of causes."

We have already noted that Newton's rule 2 is a way of increasing the independence of theory from data: to explain the motions of the moons of Jupiter and Saturn, and the Planets orbiting the Sun, all by appeal to the same sort of cause plainly results in a lower ratio of DECs to components of data explained than would explaining them by more sorts of causes.

Newton also draws a corollary to proposition V, which he uses later in his argument. The corollary is that there is a "power of gravity tending to all the Planets". He justifies this conclusion on the grounds that "doubtless, Venus, Mercury and the rest are bodies of the same sort with Jupiter and Saturn", and since they (Jupiter and Saturn) have a power of gravity, we may conclude that the other planets do also. Newton presumably sees this inference as justified by his third "rule of reasoning". However, what might seem perhaps slightly surprising about Newton's reasoning here is that he is prepared to confidently make an induction to a general conclusion on the basis of only two positive instances. It is possible that Newton is perhaps moving a little too quickly to the general conclusion in this case, but it is worth noting that there are a number of other factors which, on the framework advocated here, help to make Newton's inference rational.

First, let us remind ourselves of the circumstances in which, on the view advocated here, we make inductions. In Chapter 3 (pages 53–5) we distinguished between what we called "the neutral case", "the positive case" and "the negative case". In the neutral case, we have noted that all observed A have been found to be B, but we have no knowledge of the nature of A or of any mechanism by which a thing being an A would bring it about that it would also be a B. A case of this type exists when we have observed that a "black box", for example, hummed every time we touched it. In such a case, we would pretty quickly arrive at the hypothesis that the black box would hum every time we touched it. Moreover, on the view advocated here, this generalization would acquire a higher probability than any other generalization. The positive type of case obtains when our knowledge of A and B indicates that if some small number of observed A have been B, it is highly likely that there is a propensity for A to be B, and hence highly likely that all A are indeed B. So, on the view advocated here, in both the neutral and positive cases, only a small number of positive instances of a generalization are required for it to be rational to assert the truth of the generalization. In the negative case, our knowledge of the nature of what it is to be an A and what it is to be a B makes it unlikely that there is a propensity for A to be B. In the negative case, even a very large number of positive instances of a generalization do not succeed in making it rational to accept the generalization. So, on the view advocated here, in both the neutral case and the positive case, we are justified in accepting a generalization even if it has only a very small number of positive instances. Only in the negative case do a small number of positive instances not justify the acceptance of the generalization. But now, it is plain that the case currently before us concerning an induction from just two planets is *not* a negative case. Newton did not have any reason to believe that the nature of planets was such that it was unlikely for them to have a propensity to have a power of gravity. Hence the present case is either a case of the positive type or the neutral type. Therefore, on the view advocated here, it *is* permissible to conclude that the generalization "all planets have a power of gravity" is more likely than other generalizations on the basis of just two positive instances.

We should also note, of course, that although Newton had only two cases of objects he knew definitely to be planets with a power of gravity, he did have access to some information which suggested the Earth, too, might be a planet of the Sun. And since the Earth obviously has a "power of gravity", this would raise the number of planets with a power of gravity to three. Moreover, since he has shown that the Sun also has a power of gravity, the hypothesis is clearly suggested that perhaps all the bodies in our solar system have a power of gravity.

Further on in his first corollary to proposition V, Newton observes that it follows from his third law of motion ("to every action there is always opposed an equal reaction") that since the planets gravitate towards the Sun, the Sun must also gravitate towards the planets, and that since the Moons of a planet gravitate toward that planet, so the planet must also gravitate towards its moons. It also follows, he notes in the third corollary, that "all the planets do mutually gravitate toward one another". These claims follow deductively from conclusions already established, together with his third law of motion.

Finally, in the "Scholium" to proposition V, Newton argues that the centripetal force acting on all orbiting bodies must be gravity. He has already argued that gravity is the force keeping the Moon in its orbit. But since the orbits exhibited by the other planets are phenomena of the same sort as the orbit of the Moon, it follows that, by his first and second rules of reasoning, they are to be explained in the same way as the orbiting of the Moon, that is as due to gravity. It is plain that this hypothesis explains the orbiting behaviour of all bodies in a way that maximizes the independence of theory from data: to explain the orbit of the Moon in one way and the orbits of other bodies in another would plainly increase the ratio of DECs to components of data explained.

Proposition VI asserts: "That all bodies gravitate towards every planet, and that the weights of bodies towards any one at equal distances from the centre of the planet are proportional to the quantities of matter which they severally contain."

We will initially consider the argument Newton presents for the second part of proposition VI. First, let us remind ourselves of the conceptual distinction between the *weight* of a body and *the quantity of matter* the body contains. The weight of a body is the force it exerts "downwards", due to the gravity acting on it. This is not always the same as the quantity of matter in it: a cubic foot of lead on the Earth contains the same quantity of matter as a cubic foot of lead on the Moon, but the latter will weigh less, that is it will exert less force downwards, due the weakness of the gravity on the Moon. Newton points out that our experience shows us that, for many classes of bodies, their weights, at equal distances from the planet, are always proportional to the amounts of matter in them. In particular, he argues that this is true of three classes of bodies: (1) terrestrial bodies, that is the bodies we find around us on the surface of the Earth, (2) the moons of Jupiter and (3) the primary planets. From this he draws the conclusion that, for all bodies whatsoever, at equal distances from any planet, the weights of those bodies are proportional to the quantities of matter those bodies contain. Newton's inference from (1), (2) and (3) is simply an application of induction, which, as we have already noted, serves to increase the independence of theory from data. However, his arguments for the truth of (1), (2) and (3) are more complex, and need to be considered in some detail.

Newton says that experiments he performed with pendulums show, to a very high degree of accuracy, that the rate at which objects accelerate towards the Earth is independent of the masses of the objects, that is that objects with *different* masses all fall towards the Earth at the *same* rate. Plainly, the pendulums used in the experiments were all located at the surface of the Earth and were, therefore, all located at the same distance from the centre of the Earth. Hence, Newton concludes, objects located at the surface of the Earth all accelerate at the same rate, regardless of their masses. But now, it follows from this fact, together with the second law of motion, that, for objects located at the surface of the Earth, the force acting on them, causing them to fall, is always proportional to the "quantity of matter" they contain.

In discussing the results of experiments that he performed with pendulums, he says, "I could manifestly have discovered a difference of matter less than the thousandth part of the whole, had any such been" (Newton 1960: 411). So we may say that Newton had satisfied himself that, at the surface of the Earth, the force acting on a body is proportional to its mass to an accuracy of at least one part in one thousand. From this he concludes that the force is, in fact, *exactly* proportional to mass. There are a number of ways in which it is possible, within the general framework advocated here, to account for this inference. First let us recall that if mass, or quantity of matter, is the only aspect of an object affecting its weight, there is a close-to-*a priori* argument showing that weight will be proportional to mass. Consider two objects of the same mass M, such as two bricks. At the surface of the Earth, each brick pushes down with a force F. Plainly, then, if the two bricks were tied together, or simply regarded as a single object, each of them would continue to push down with a force F, since presumably the act of tying them together would not affect this force. But then the compound object consisting of the two bricks, which has a mass of 2M, will push down with a force of 2F. The argument can plainly be generalized to show that *if* mass is the only factor affecting weight, then weight will be proportional to mass. But perhaps mass is not the only factor affecting weight – perhaps colour, or shape, or temperature also have an influence. But now, the observations available to Newton concerning objects at the surface of the Earth are compatible with mass being the only influence. Therefore, the explanation with the fewest DECs compatible with the data, and therefore the explanation with the highest degree of independence from the data, will be the one that says that mass is indeed the only factor influencing weight. But it is an *a priori* plausible consequence of this explanation that weight is *precisely* proportional to mass. So if we aim to maximize independence, it is rational to conclude that weight is indeed precisely proportional to mass for objects at the surface of the Earth.

Newton then (1960: 412) proceeds to argue that weight is proportional to mass for the moons of Jupiter.[14] He observes that, according to phenomenon I, the moons of Jupiter conform to the analogue of Kepler's third law of motion, and that from results already established it follows that the rate at which they accelerate towards Jupiter must vary as the inverse square of the distance from Jupiter's centre. From this it trivially follows that, at equal distances from Jupiter's centre, different moons with different masses will accelerate towards Jupiter at the same rate. Therefore, the forces acting on the moons must be proportional to the masses of those moons. Consequently, Newton concludes, that at equal distances from Jupiter the gravitational force acting on the moons is always proportional to the masses of the moons. As these results would appear to follow deductively from results already established, we need not concern ourselves here with the nature of Newton's reasoning. Essentially the same type of reasoning is used by Newton in establishing that, at equal distances from the Sun, the gravitational force exerted by the Sun on the primary planets is proportional to the masses of those planets.

Let us now turn our attention to Newton's argument for the first part of proposition VI, that is that all bodies gravitate towards every planet. In arguing for this claim, Newton makes use of corollary 3 of proposition V: that all the *planets* mutually gravitate towards one another. He argues that it follows from the fact each *planet* gravitates towards every other planet that each *body* gravitates towards every planet. We can, I think, assume that by a "body" Newton means a "part of a planet". Let us suppose, contrary to proposition VI, that some body (that is some part of some planet) did not gravitate towards the other planets even though that body had *mass*. We will, that is, suppose that although the body had *mass*, it was "gravitationally inert". Then, Newton reasons, the planet containing that body would not gravitate towards the other planets in proportion to its mass; it would, rather, exert a gravitational influence on the other planets somewhat less than its mass. But since it has been established that all planets do gravitate towards all other planets in proportion to their mass, it follows that no planet contains a body that is gravitationally inert. Newton then generalizes this argument. He writes:

> The weights of all the parts of every Planet are one to another as the matter in the several parts. For if some parts did gravitate more, others less, than for the quantity of their matter, then the whole Planet, according to the sort of parts with which it most abounds, would gravitate more or less than in proportion to the quantity of matter in the whole.
>
> (Newton 1960: 413)

On the face of it, it seems to be possible to object to Newton's argument at this point.[3] It surely seems to be possible for a planet to be composed of gravitationally strong and gravitationally weak parts in such a way that their influences "cancel" each other out, with the gravitational force of the planet as a whole coming out as proportional to its mass. However, it will be argued that Newton was right not to accept that this possibility was actually the case, and that Newton's position can be explained within the view advocated here.

Let us suppose that one or more of the planets were composed of gravitationally "mixed" components, and they just happened to cancel each other out in such a way that the gravity of the planet was exactly proportional to its mass. This would surely be a highly *a priori* improbable state of affairs: it seems much more likely that one sort of matter would be more common than the others, in which case the gravitational power of the planet would *not* be exactly proportional to its mass. But Newton has established that, for all planets, their gravitational power is exactly proportional to their mass, from which result it is rational to conclude that it is most likely that, for each part of a planet, its gravitational power is exactly proportional to its mass. Of course, it might be possible to provide some explanation of why each planet happened to be composed of the right mixture of gravitationally different types of matter. Call this explanation H. But then Newton's system would have to be supplemented with hypothesis H if it was to explain all available data. It would therefore have more DECs, and so a higher ratio of DECs to data, than the system Newton actually advocates. Consequently, it would have a lower degree of independence from the data than Newton's actual system. Hence, we see Newton's choice of hypotheses as maximizing the independence of theory from data.

Proposition VII is: "That there is a power of gravity tending to all bodies, proportional to the several quantities of matter which they contain." This follows deductively from results already established. In corollary 3 to proposition V it was established that all the planets mutually gravitate towards one another, and that this gravitational force varies as the inverse of the square of the distance. From this, together with proposition XLIX of Book I, it follows that the gravitational force directed towards each planet is, in every case, proportional to the mass of the planet. Moreover, in proposition VI it was established that all *bodies* gravitate towards every planet, and the extent to which they do is proportional to their masses. But now, it follows from these results, and from Newton's third law of motion (for every action there is always opposed an equal reaction), that each planet will gravitate towards every body, and the extent to which it does this will be proportional to the mass of the body. That is, that there is a "power of gravity tending to all bodies, proportional to the several quantities of matter which they contain". Since proposition VI follows deductively from results already established, we need not here concern ourselves with this final step in his reasoning.

SUMMARY OF CONCLUSIONS SO FAR

So far it has been argued that the case Newton makes for his three laws of motion can be seen as applications of the arguments advocated earlier in this book. His "rules of reasoning in philosophy" can all be seen as strategies for increasing the independence of theory from data. The case he makes for the six "phenomena" on which his argument for universal gravitation are based can also be seen as applications of the principles and arguments advocated here, as can his argument for universal gravitation itself. We therefore have a probabilistic explanation of the subsequent empirical success of Newton's theories. We have also given an account of how it is that Newton's laws should seem to be close to *a priori* and yet enjoy surprising subsequent empirical confirmation.

NEWTON AND THE SECOND OF THE PHENOMENA WITH WHICH WE ARE HERE CONCERNED: KNOWLEDGE OF LESS ACCESSIBLE PARTS OF REALITY

As noted in the first chapter, sometimes scientists give us descriptions of parts of the world at the time inaccessible, and those descriptions subsequently turn out to be correct. This was the second of the phenomena which it is the aim of this book to explain. One example we have already mentioned is the postulation of the planet Neptune by Adams and LeVerrier and the subsequent verification of the literal reality of that planet by telescopes and spacecraft. In this section it will be argued the results established in this chapter, and in earlier chapters, make it possible for us to explain at least one case of this phenomenon.

At first sight it might be thought that the view advocated in this book would be unable to account for this type of success. In Chapter 4 it was argued that the predictive success of science can be explained without asserting the reality of the theoretical entities postulated by the successful theory. The explanation used appealed to a particular type of pattern in the data: although the characterization of the pattern might employ terms that putatively refer to unobservables, in order to explain the ability of the pattern to lead to successful predictions, we only needed to appeal to the *a priori* improbability of *the pattern*; we did not need to make the additional claim that the putatively referring terms in it really were referring. The hypothesis of scientific realism was otiose to the explanation. Since the explanation of success used here bypasses the question of the reality of the entities postulated by successful theories, it might be thought that it would be unable to explain why it is that sometimes the existence-claims made by scientific theories do turn out to be correct. But this is not so. It will be argued that if we attend closely *to what it is that actually requires explanation* in these cases, an explanation of them follows quite simply from this account.

Consider the case of Neptune. We are now entirely confident that the planet Neptune is not merely an entity of theory. Even the strictest instrumentalist must surely agree that it is as real as a table or a rock. Perhaps, among all the forms of evidence we have for its reality, the most uncontroversially compelling is the simple fact that spacecraft have flown past it, taking photographs of a large, predominantly blue sphere consisting largely of hydrogen and methane. Of course, observations of this *precise* nature were not necessary to persuade us it was real: it could, for example, have been some shape other than a sphere, some colour other than blue and composed of different chemical substances. But if a photograph is taken of an object with a definite shape and colour, and which is made of known chemical substances, then even the most extreme Instrumentalist would agree there was a real object there. And so I think we may say that the phenomenon that requires explanation is this: how is that Newton, together with Adams and LeVerrier, were able to arrive at a theory which postulated an entity which was not accessible at the time, but which did turn out to have some features we regard as indicative of full-blown reality, such as shape, colour and chemical constitution. The answer to be given is along the following lines: there was available, prior to the era of space travel – and even available to Newton – data which permitted the formation of some additional hypotheses H. These additional hypotheses entailed that there would be, beyond the orbit of Uranus, an entity that had shape and mass. But if it had mass, it must be made of matter and so, plausibly, must have colour and chemical make-up. Moreover, these additional hypotheses were, it will be argued, highly independent of the data. So, on the view advocated here, they were likely to lead to correct predictions. This provides us with an explanation of why data confirmatory of an object with a shape, a colour and a chemical constitution was indeed subsequently obtained.

It was, of course, a prediction of Newton's theory, as worked out by Adams and LeVerrier, that there would be a (large) *mass* at a particular location in space, beyond the orbit of Uranus. Let us refer to this region as R. It is not a consequence of the view advocated here that it is likely that there *should actually be* such a mass in region R. But since – it has been argued – Newton's theories are highly inde-pendent of the data, it *does* follow on the account given here that it is likely that *data* indicative of a mass in region R would be obtained. So let us accept that there follows from this view a statement of the form "If observations were made of region R, data would be obtained indicative of the presence of a mass in R." But now it is also possible to make a number of inductive generalizations from past obser-vations of masses. Among these inductive generalizations are the following two:

1. Whenever observations have been made indicative of the presence of a mass in any region S, provided the mass has been large enough, there has also been evidence in region S of either a solid or a liquid or a gas.
2. Whenever observations have been made indicative of the presence of a mass in a region S, provided the mass has been large enough, there has also been evidence of the presence in region S of something that is either transparent, or that has colour(s).

Clearly enough, these generalizations enable us to draw the conclusion that orbit-ing in region R beyond the orbit of Uranus there will be found evidence of an object that is either solid or liquid or gas, and which is either transparent or has colours.

Now, there are two other inductive generalizations that could have been made before the period of space travel. These are:

3. All large objects orbiting the Sun are approximately spherical; and
4. All objects orbiting the Sun reflect their light from the Sun.

From the second of these generalizations it follows that there will be found in R evidence of something that is *not* transparent and which therefore has colour(s). And from the first it follows that it will be approximately spherical. Hence, in conclusion, it is possible to derive from the assertion "if observations were made of region R, data would be obtained indicative of the presence of a mass in R", together with other inductively confirmed generalizations, the conclusion that "if observations were made of region R, data would be obtained indicative of the presence of an object in R that is approximately spherical, that has colour(s), and is either solid, liquid or gas". By the argument given in Chapter 3, these inductive generalizations are likely to lead us to correct predictions about the data. But this provides us with an explanation of the phenomenon with which we are presently concerned: it explains why it was that when a spacecraft travelled to a region beyond the orbit of Uranus, it obtained photographs of an object with a definite shape, colour and chemical composition. And this is sufficient to explain this case of phenomenon 2.

Although the explanation of success given here does not presuppose scientific realism, the notion of the independence of theory from data can, under certain conditions, come very close to justifying a scientific realist view of some entities. It can be used to justify the claim that we can expect observations confirming certain entities that have size, mass, shape, colour, chemical composition and so on. It could, in principle, justify the claim that some postulated entities are *observationally indistinguishable* from real entities. It can thus, perhaps, come "asymptotically close" to justifying a fully realist view of some entities.

CHAPTER 7
Special relativity

INTRODUCTORY REMARKS

In this chapter we will consider Einstein's arguments for special relativity. It will be argued that the various inferential steps taken by Einstein in developing the theory can be explained within the framework developed here.

It is highly desirable for the view offered here to be able to account for special relativity. The theory has led to some very surprising predictions, which have received subsequent confirmation. One of these surprising predictions is the phenomenon of "time-dilation", or the slowing of time in an object moving very quickly with respect to the observer. It has, for example, been experimentally observed that certain particles accelerated close to the speed of light have their normally very brief life-spans greatly extended, consistent with the time-dilation predicted by the special theory. Another novel prediction of the theory, the "twin paradox", was confirmed when it was found that when two very accurate clocks were moved in different ways and then reunited, there was a noticeable difference between the time registered by them (Hafele & Keating 1972: 166–70). Again, the difference was consistent with that predicted by relativity. Or again, the special theory of relativity leads to the famous equation $E = mc^2$, which in turn has a role to play in predicting atomic explosions, nuclear fusion, the energy released in atomic reactors and the loss of mass accompanying such phenomena.

It can hardly be denied that these examples of novel predictive success are very impressive. From the point of view of common sense, the claim that time will run more slowly the faster that an object moves, or that a moved object will age more slowly than one that remains stationary, are perhaps some of the most counter-intuitive predictions ever made. And they were certainly *novel* predictions: Einstein had not observed time-dilation, clock retardation produced by motion or mass loss due to the release of energy when he first formulated the theory in 1905. So it is desirable that the account offered here be able to explain these examples of novel predictive success.

Like the law of universal gravitation, the special theory of relativity is supported by a long and intricate series of arguments. Here it will be argued that the (non-deductive) inferences made by Einstein in developing these arguments either serve to increase the independence of theory from data or are explicable as AIM inferences.

One of our aims here is to explain how scientists have managed to hit upon theories that have enjoyed the types of success described in the first chapter. So if we are to explain how Einstein arrived at the Special Theory, we must turn our attention to the reasoning process he himself actually went through in arriving it. Einstein (1920) has told us that he has given an account of these reasoning processes "on the whole, in the sequence and connection in which they actually originated" in his popular book *Relativity: The Special and General Theories* (hereafter "*Relativity*"). Accordingly, here we will mostly follow the exposition given by Einstein in that book, although we will occasionally refer to other writings. The discussion of Special Relativity given below does not presuppose any prior familiarity with the theory.

EINSTEIN'S VERIFICATIONISM AND CONCEPT EMPIRICISM

At the beginning of *Relativity*, Einstein makes some remarks concerning the meaning of the term "true". He writes:

> The concept "true" does not tally with the assertions of pure geometry, because by the word "true" we are eventually in the habit of designating always the correspondence with a "real" object; geometry, however, is not concerned with the relation of the ideas involved in it to objects of experience, but only with the logical connection of those ideas among themselves. (Einstein 1920: 2)

It is worth noting that in the above quote, Einstein uses the word "real" in "scare quotes". Later in the same sentence he says pure geometry is not concerned with the "objects of experience", suggesting that by a "real" object, Einstein means an "object of experience". But it appears that by "experience", Einstein does not necessarily mean something like "direct sense experience" or "sense-data". He seems to generally have in mind ordinary states of affairs we can perceive with the (unaided) senses, such as books on tables and the like, and the directly observable results of experiments and measurements. A statement is said to be "true", in Einstein's view, if and only if what it says corresponds with what we can observe to be the case with observable phenomena, in particular with the observable results of experiments and measurements. We may therefore attribute to Einstein a sort of verificationism.[1]

Einstein also seems to subscribe to a view that could fairly be called "concept empiricism". By "concept empiricism" I mean the idea that any meaning that a term has must be capable of being analysed in terms that refer only to experience, where, again, "experience" refers to directly observable states of affairs such as the results of experiments and measurements. For example, on page nine of *Relativity*, he writes, "We must entirely shun the vague word 'space', of which, we must honestly acknowledge, we cannot form the slightest conception." The notion of "motion through space" is to be replaced by the notion of "motion relative to a practically rigid body of reference", and the notion of distance by that of "two marks on a rigid body". Similarly, time is to be defined in such a way that "time-values can be regarded as essentially as magnitudes (results of measurements) capable of observation". This can be done if we stipulate that two episodes take up an equal amount of time if and only if they are measured as equal by two clocks of identical construction. More generally, any statements attributing spatial or temporal properties to things must, on Einstein's view, be capable of being analysed as the results of actual or possible observations or measurements. Einstein's verificationist view of truth can perhaps be seen as a result of his concept empiricism applied to the term "true".

As we will see, Einstein's concept empiricism enters into the arguments he presents for special relativity. This raises a question for the point of view defended in this book. One aim of this book is to provide an explanation of some important cases of novel predictive success. More specifically, it is to argue that the various ideas defended here: the notions of the independence of theory from data and the AIM inference, are sufficient to explain novel success. But we have just noted that Einstein apparently accepts something like concept empiricism and a form of verificationism, and that these doctrines apparently enter non-redundantly into his arguments for special relativity. The question therefore arises whether this can be accommodated into the perspective to be defended here. This is an issue to which we will return periodically throughout this chapter. But, for the moment, we simply need to note that Einstein's concept empiricism has two consequences:

1. Any statement attributing spatial or temporal properties to anything must be capable of being analysed as the results of actual or possible results of measurements; and
2. The notions of absolute space and time are to be regarded as meaningless (unless it should prove possible to give an analysis of these notions in terms of the results of actual or possible observations or measurements).

THE PRINCIPLE OF RELATIVITY

Einstein's principle of relativity is as follows:

> Let K and K* be two frames of reference. Then, if relative to K, K* is inertial and non-rotating, then natural phenomena run their course with respect to K* according to exactly the same general laws as with respect to K.
>
> (Einstein 1920: 13)

Einstein's seems to have a number of reasons for accepting the principle of relativity.[2] One of these is straightforwardly empirical. He notes (Einstein 1920: 15) that the Earth is rotating around the Sun at about 30 kilometres per second, and that the direction of its motion is changing throughout the course of a single year. The motion of a frame of reference that is fixed relative to the Earth is therefore constantly changing. Let us refer to as "K" a frame of reference that is fixed relative to the Earth on a particular date, say 1 January. Then there will be a series of other frames of reference, K*, K** and so on, that are fixed relative to the Earth on other dates. These other frames will be, as near as does not matter, inertial with respect to K. But, Einstein notes, it is a well confirmed fact that the general laws obeyed by the objects on the Earth do not change throughout the course of a year. This is so not just for Newton's laws of motion, but for the laws of electromagnetism as well. That is, the laws that objects obey with respect to K are the very same laws as those they obey with respect to K*, K** and so on. Einstein sees this as a "very powerful argument" in favour of the conclusion that if objects obey certain laws with respect to some frame K, they obey the same laws with respect to *any* frame that is inertial and non-rotating with respect to K, that is for the principle of relativity.

Einstein's inference from the unchanging character of natural laws throughout the year to the principle of relativity can be seen an application of *Newton's* second and third rules of reasoning. Assuming it to be the case that any frame of reference stationary with respect to the Earth is as near as does not matter to an inertial frame, we can say that, for all *inertial* frames within the reach of our experiments, the laws obeyed by objects are the same. But we are not entitled to draw this conclusion for *all* frames, *whether accelerated or not*, since the property of being accelerated is one that does "admit of intensification or remission of degrees". And, as we noted in the previous chapter, the application of Newton's rules of reasoning serves to increase the independence of theory from data. So, Einstein's acceptance of the principle of relativity is explicable within the point of view developed here.

It should be noted, however, that Einstein also gives another argument for the principle of relativity that is a little more complicated than the one just mentioned. He asks us to suppose that one particular co-ordinate system – call it K – is really at rest, and the others are really in motion. In particular, let us suppose that on a particular date, say 1 January, the Earth is "really" at rest, and so a co-ordinate system at rest with respect to the Earth on that date is "really" at rest. On other dates, the Earth, and any "attached" co-ordinate system, is really in motion. Then on all dates other than 1 January, there will be a factor present in all the phenomena that take place on the Earth that is not present on 1 January, and that factor is that the phenomena are *passing through* absolute space, or *passing through* the ether. If this additional factor is present on the other days, then on those other days the phenomena on Earth will be produced by a more complicated mechanism than on 1 January, when the Earth is not passing through space. And so we would expect the *explanations* of the phenomena on Earth to be at their simplest on 1 January and more complicated on other days. But, Einstein points out, no such variation in the explanations required as the Earth moves around the Sun has ever been observed. He says "the most careful observations have never revealed [such variation]". The hypothesis that the Earth is moving relative to absolute space therefore proves to be explanatorily idle.

We should observe that Einstein has available yet another, quite independent reason for expecting the principle of relativity to be true. Suppose it were not true, that is, that the laws things obeyed in one frame were different from those they obeyed in another. We would naturally seek an explanation of this fact. If the only apparent *difference* is that the two frames were moving at different velocities, then the explanation of the difference in the laws would, it seems, have to appeal to the difference in velocities of the frames. But then the question would arise "With respect to what are the two frames moving at different velocities?" The answer to this question would, presumably, have to be "with respect to absolute space". But Einstein has already asserted that "we can form no conception whatsoever" of the notion of absolute space. And if the notion of absolute space is meaningless, then the claim that the two frames are moving at different velocities with respect to absolute space would also be meaningless. Moreover, if the claim is *meaningless*, then, trivially, it cannot be true. This would lead us to expect that there should be no difference in the laws obeyed by objects relative to the two frames, since any such difference would have to be due to "nothing at all", or to something of which we can form no conception. So Einstein's concept empiricism also leads us to expect that the principle of relativity ought to be true.

We see here a case of two independent arguments both leading us to the same conclusion, namely the truth of the principle of relativity. The concept empiricist argument leads us to say that the principle should be *precisely* correct. Observations made of objects on the surface of the Earth at different times of the year confirm that the principle is correct to within the accuracy of our observations. So, by the argument from the agreement of independent methods, we may draw the conclusion that the principle of relativity is precisely correct.

It is also worth noting that the *empirical* evidence for the principle of relativity does not appear to be Einstein's only reason for accepting it. On page 19 of *Relativity* he writes, "We should retain the principle of relativity, which appeals so convincingly to the intellect because it is so *natural* and *simple*" [my italics]. First, it is intuitively clear that the principle of relativity is *simpler* than any alternative which said that the laws which things obey relative to K* are different from those they obey relative to K. Einstein does not say why the principle is "natural", but it is possible that what he has in mind is that if the laws *were* different in different frames, this difference would have to be due to motion with respect to absolute space, that is with respect to "nothing at all". It is also possible that he sees it as natural because it postulates a "symmetry" in the technical sense of that term: the principle of relativity tells us that the laws of nature remain unchanged if we change our (inertial, non-rotating) frame to any other (inertial, non-rotating) frame. But in Chapter 5 (pages 114–16), it was argued that a preference for symmetrical theories can be explained on the view advocated here.

THE MICHELSON–MORLEY EXPERIMENT

The principle of relativity says that all the laws are the same in all frames inertial with respect to each other. The question arises whether Einstein includes among these laws the law of the constancy of the speed of light. The answer to this question is given unambiguously just a few pages later in *Relativity* (Einstein 1920: 18), when he writes, "Like every other general law of nature, the law of the transmission of light *in vacuo* must, according to the principle of relativity, be the same for [a moving object such as a] railway carriage as reference body as when the rails are the body of reference."

A few pages earlier Einstein had said that a "very powerful" argument in favour of the principle of relativity was the fact that the laws of nature had not been observed to change as the Earth moves around the Sun. That he is prepared to include the constancy of the velocity of light among these laws is, presumably, due to his awareness of the Michelson–Morley result, although Einstein does not at this point mention that result by name.

It is worth describing the Michelson–Morley experiment in some detail, as it has a number of features that will be important later. At the time of this experiment (1887) it was believed that light was a wave. But in all known cases of other wave phenomena – water waves, sound waves, waves transmitted along stretched strings and so on – there was always a medium through or along which the wave travelled. Through what medium did light waves travel? Although it was not known what this medium was, it was thought that it must exist. It was referred to as the "lumeniferous ether" or just as "the ether".

The hypothesis that light is transmitted through the ether gives rise to a testable prediction. As we have noted, the Earth is rotating around the Sun at about 30 kilometres per second. If light is travelling through the ether, and the Earth is also travelling through the ether, then the velocity of light *relative to the Earth* ought to change appreciably over a period of six months, since during that period the velocity of the Earth relative to the ether will have changed by about 60 kilometres per second. The Michelson–Morley apparatus was designed to detect this predicted change in the velocity of light as the Earth moves through the ether.

The Michelson–Morley apparatus made use of the following fact. Suppose that on a particular date (say 1 January) the Earth happened to be at rest with respect to the ether. Then a beam of light shot out in, say, a northerly direction would take exactly the same time to travel one metre as a beam shot out in a westerly direction. Similarly, a beam shot out in a northerly direction for a distance of one metre, and then reflected back to its source, would take exactly the same amount of time to get back to its source as a beam shot out in a westerly direction and reflected back to its source. But if the Earth is in motion with respect to the ether, this need not be the case: the two beams can take different times to return to their source since the two beams will always be travelling at the same velocity *with respect to the ether*. Michelson and Morley noted that it would be possible to tell whether or not the two beams returned to the source simultaneously by observing whether or not the two beams were *in phase* when they returned. If the two beams were in phase – that is, if the peaks and troughs of the waves making up the two beams exactly coincided – when they returned to the source, it was reasonable to suppose that they had taken exactly the same time to make their journeys, and hence that the velocity of light was the same in both directions. But if the two beams were out of phase, they must have taken different times, and hence that the velocities of the two beams relative to the apparatus, and hence relative to the Earth, must be different. This led to the prediction that if light was indeed travelling through the ether, then, as the Earth changed its motion through the ether over a six month period, the beams should move in and out of phase, and that the extent to which they were out of phase also ought to change. But when the experiment was performed, it was found that the two beams were always in phase, no matter in what direction the Earth was travelling. This was taken to show that the velocity of light was always the same, no matter at which velocity the observer was travelling. This constant velocity of light is traditionally denoted as "c".

There are two aspects of the Michelson–Morley experiment that we need to note:

1. The Michelson–Morley apparatus measures the velocity of light by measuring the time it takes for a beam to travel from its point of origin A, be reflected from a point B and then return to A; that is, it measures the total time it takes for it to get from A and back to A again. The Michelson–Morley apparatus does not show how long it takes light to get from A to B. So the apparatus only tells us the average *two-way* velocity of light. It does not measure the "one-way" velocity of light.
2. The evidence for the claim that the velocity of light is the same in both directions is that the two beams shot out at right angles always return to their point of origin *in phase*.

Both of these aspects will become important in later discussion.

THE PROBLEM MOTIVATING EINSTEIN TO DEVELOP
THE SPECIAL THEORY

In *Relativity* Einstein says that problem which motivated the development of the theory is an apparent contradiction that arises between the law of the constancy of the speed of light and some eminently reasonable – and in fact apparently undeniable – assumptions concerning the behaviour of relative velocities (Einstein 1920: 17–20). Let us imagine an observer O measuring the speed of a beam of light directed towards some distant object A. We assume O to find the velocity of the beam to be c. We now imagine another observer O* moving towards the distant object A in a straight line at constant velocity v relative to O, where v is greater than zero although less than c. Common sense tells us that if the beam of light is travelling at c relative to observer O, and O* is travelling in the same direction as the beam at velocity v, then the velocity w of the beam of light relative to O* will be given by:

$$w = c - v \qquad\qquad\qquad (1)$$

As (1) is the formula that would be used in classical (Newtonian) physics to calculate the velocity of light relative to O*, we will refer to it as (an instance of) the *classical principle of the composition of velocities*. Plainly, if v is greater than zero, then, necessarily, w must be less than c. But this contradicts the principle of relativity, which asserts that all laws, including the law of the constancy of the speed of light, must be the same for all inertial frames. The principle of relativity asserts that if O* were to measure the speed of light, the result would be the same as that obtained by O, that is the velocity c. But if O* were to find the speed of light to be c, then common sense would lead us to expect O to find the speed of light to be $w = c + v$, again contradicting the principle of relativity. The principle of relativity therefore contradicts the classical principle of the composition of velocities. In *Relativity*, Einstein sees his problem as that of removing this contradiction.

The initial reaction of naïve common sense is, surely, to assume that it is the principle of relativity that must be wrong, since the classical principle of composition of velocities given in (1) seems so obvious. But Einstein argues that the principle of relativity ought to be retained: it is the classical principle of the composition of velocities which he says ought to be rejected.

Einstein's concept empiricism plays a crucial role in his decision to reject the classical principle. We recall that his concept empiricism is the doctrine that any assertion whatsoever, including expressions such as "event E occurred at time t", "object O is X metres long" or "object O is moving at velocity v" must be capable of being analysed as the results of actual or possible observations or measurements. *Einstein's central claim, the core of the solution to his problem, is that when expressions such as these are given their most "natural" definitions in terms of the results of possible measurements, the classical principle given in (1) turns out not to be correct.* He begins his development of this idea by discussing concepts such as "space", "time", "rest" and "motion" from a concept empiricist standpoint.

EINSTEIN'S REJECTION OF THE DISTINCTION BETWEEN ABSOLUTE REST AND MOTION

Einstein appears to have two distinct reasons for rejecting the distinction between absolute rest and motion. The first comes from his rejection of absolute space. To assert that an object is in a state of absolute rest is, presumably, to assert that it is at rest with respect to absolute space. But Einstein has already claimed that we cannot form any conception of "absolute space", that the notion is without any meaning whatsoever. And his rejection of the meaningfulness of this notion is a consequence of his concept empiricism. So, for Einstein, the notion of absolute rest is also without meaning.

However, he also has another reason for rejecting the concepts of absolute rest and of motion with respect to absolute space. Suppose the concept of absolute space was meaningful. Then, as the Earth moved around the Sun, its motion with respect to absolute space would be constantly changing: if, for example, it was at rest with respect to absolute space on a particular date, such as 1 January, it would be moving with respect to absolute space on later dates, and its motion would be constantly changing. But if it is to be rational to believe this, there must be some *empirical* evidence that the motion of the Earth is changing with respect to absolute space. Its changing motion relative to other *objects*, such as the Sun, plainly does not constitute such evidence, since it is only evidence for motion relative to other objects, not to "space itself". The only other type of empirical evidence that there could be, it seems, would be a change in the laws that objects obeyed as we changed from one frame of reference to another. But the principle of relativity tells us there is no such change. And the correctness of this principle is supported by the fact that the laws obeyed by objects do not change as the Earth moves around the Sun. Therefore, there is no empirical evidence for motion relative to "space itself". The hypotheses that there are such things as "being at rest with respect to absolute space" and "being at motion with respect to absolute space" are hypotheses without any form of empirical support. For Einstein, therefore, the distinction between absolute rest and motion is explanatorily otiose and so ought to be rejected. This is plainly in keeping with maximizing the independence of theory from data. The hypothesis "there is a distinction between absolute rest and absolute motion" has no role in explaining any data. Therefore incorporating such a hypothesis into our total theory increases the ratio of DECs to components of data explained, and so decreases the independence of theory from data.

We see here another case of two distinct arguments leading us to the same conclusion. First, Einstein's concept empiricism leads him to reject the concept of absolute rest, and of the distinction between absolute rest and absolute motion. But the principle of relativity, and the empirical evidence for it, also supports rejecting the distinction between absolute rest and motion. Einstein's rejection of the distinction between absolute rest and motion can therefore be seen as an application of the AIM inference.

Much the same arguments can be used against the idea of absolute time. As we have already noted, Einstein has said any reference to, for example, the passage of a quantity of time must be capable of being analysed as the result of actual or possible measurements. Einstein's concept empiricism leads him to reject any concept of absolute time beyond that which is measured by accurate clocks. Moreover – although Einstein himself does not say this – we do not note any change in the laws obeyed by things with the passage of time. Therefore, the considerations which tell against absolute space would also seem to equally tell against absolute time, and the rejection of both notions can be seen as AIM inferences.

EINSTEIN'S DEFINITION OF THE SIMULTANEITY OF
DISTANT EVENTS

As we have already noted, Einstein's concept empiricism forbids him from taking an expression such as "distant events A and B occurred at the same time" as an unanalysed primitive. It is, rather, to be analysed in terms of the results of possible observations or measurements. The definition he proposes is as follows:

> Suppose M to be a point mid-way between the location at which A occurred and the point at which B occurred. Then, if the light from A and the light from B arrive at point M simultaneously, then events A and B are said to be simultaneous. (Einstein 1920: 23)

This defines distant simultaneity in terms of local simultaneity, but the light from A and from B can be said to arrive at M at the same time if an observer located at M perceives the light from A and B at the same, the notion of "perceiving at the same time" apparently being acceptably clear from a concept empiricist standpoint.

One apparent difficulty with this definition of simultaneity is that it assumes that the light travelling from A to M and the light travelling from B to M do so at exactly the same velocity. If, for example, the light travelling from A to M moved more slowly than the other, the simultaneous arrival of the two light beams at M, would, we would conclude, show that the events A and B were *not* simultaneous. Einstein's response to this objection is, on the face of it, rather curious. He writes, "That light requires the same time to traverse the path A → M as for the path B → M is in reality neither a *supposition nor a hypothesis* about the physical nature of light, but a *stipulation* which I can make of my own freewill in order to arrive at a definition of simultaneity" (1920: 23).

Some features of Einstein's position here might seem rather puzzling. He writes, "There is only *one* demand to be made of the definition of simultaneity, namely that in every real case it must supply us with an empirical decision as to whether the conception that has to be defined is fulfilled" (Einstein 1920: 23). To be sure, from a concept empiricist stand point, this is a condition that must be fulfilled by any definition of anything, but it would not appear to be sufficient to ensure that what has been defined is *simultaneity*. Suppose I say that if one car travelling at 10 kph leaves event A and another travelling at 100 kph leaves event B and the two cars meet at the mid-point M simultaneously, then the events A and B are said to be *schimultaneous*. Then, from a common-sense point of view, whatever "schimul-taneous" means, it does not mean "simultaneous". Or again, if I were to say that I stipulate that two cars take the same *schtime* to make their journeys, whatever meaning the term "schtime" has acquired by this stipulation, we would not ordinarily say it has come to mean "time".

I suspect that Einstein would respond that in the case of the cars we have other means of verifying, and therefore attaching a meaning to, for example, the claim that the two cars took different times to make their journeys, whereas in the case of light travelling from A to M and from B to M, we do not have any means of verifying either that light took the same time to make the two journeys or that it took two different times. Therefore, in this case, and granting Einstein's concept empiricism, neither the claim that the journeys took the same "time" nor that they took different "times" has any meaning, on the ordinary meaning of "time". Einstein's stipulative definitions of "time" and "simultaneity" are, therefore, within this view not in any way contrary to or incompatible with the ordinary meanings of those terms; they are, rather, *extensions* or *additions* to the ordinary meanings of the terms. They are, roughly speaking, no more incompatible with the ordinary meanings of "time" or "simultaneity" than "'p and q' is incompatible with 'p'".

But even if we grant that Einstein's definition of "simultaneity" is *compatible* with the ordinary meaning of that term, another question arises: "Why extend the ordinary meaning of "simultaneity" in the way that Einstein advocates; why not extend it in some different way?" Here Einstein says (1920: 27) that his definition is the most *natural* definition of simultaneity. While he does not explain what he means by "natural", there are a number of things that he might have in mind. First, it is clear that any alternative definition would have an element of arbitrariness not present in Einstein's definition. Suppose it were asserted that the velocity of light from A to M was in some way different from its velocity from B to M. Then there would seem to be an element of arbitrariness in the choice of any particular ratio of the two velocities that, intuitively, does not seem to be present in saying that the velocities are the same. Suppose, for example, it is said the velocity from A to M is 1.1 times greater than the velocity from B to M. But then the question would surely arise: "Why 1.1 rather than 1.11, or 1.2, or 5.3425?" Moreover, we know from the Michelson–Morley experiment that the average two way velocity of light remains constant. So if it is asserted that the velocity of light changes with direction, it must also be asserted that it changes in such a way that the *average* of its velocities in opposite directions always remains the same. Plainly, this constancy of the *average* of velocity could not simply be a fluke, and so would require explanation.

These considerations help us, I think, to get a grip on what is behind Einstein's suggestion that his definition of simultaneity is the most *natural* one. In the absence of any reason why the velocities ought to be different, it is most natural to assume that they are not different. It is also plausible to think that the only thing that *could* constitute a reason why the velocities would be different would be if A and B were in motion with respect to absolute space, or with respect to the ether. But since, Einstein has accepted, there is no such thing as absolute space or the ether, we would have no *explanation* of why the velocities in the two directions should be different. Neither would we have an explanation of the *a priori* improbable fact that, despite the fact that the speed changes with direction, the average two-way velocity always comes out the same. But saying that the velocity is the same in all directions provides an extremely simple explanation of why the average two-way velocity should always be the same. Consequently, saying that the velocity of light has one velocity in one direction and another velocity in the opposite direction, and that these different velocities always average out to the same velocity, is clearly to postulate more theoretical regularities than to say it only has one velocity. Therefore, saying that velocity changes with direction gives rise, within our total theory, to a higher ratio of DECs to data explained and therefore to a lower degree of independence than does saying velocity remains the same. Moreover, to assert that light has one velocity in one direction and another in a different direction gives rise to an increase in dependence that seems likely, from Einstein's point of view, to be ineliminable. We could eliminate this increase in dependence by producing a single, deeper theoretical regularity that explained why light should have one velocity in one direction, and another velocity in another direction. But it is hard to see how any such explanation could be given without appealing to some notion such as absolute space or the ether. But since, for Einstein, such entities do not exist, the prospects for eliminating the increase in independence in this way seem slight. The hypothesis that the velocity of light is the same in both directions is therefore the hypothesis that has the best chance of maximizing the independence from data.

EINSTEIN'S ARGUMENT FOR THE RELATIVITY OF SIMULTANEITY

It is worthwhile at this point to summarize the main conclusions at which Einstein has arrived so far. These are:

1. That absolute space and absolute time do not exist.
2. That there does not exist any difference between absolute rest and absolute motion.
3. The velocity of light is finite.
4. The principle of relativity, that is the principle that if objects obey laws L when measured from one frame of reference K, they will obey those very same laws when observed from any other frame inertial with respect to K. These laws are taken by Einstein to include the law of the constancy of the ("two-way" average) speed of light.
5. That simultaneity is to be defined as follows: event A and event B are said to occur simultaneously if and only if light from A and light from B arrive simultaneously at a point M located mid-way between A and B.
6. It is to be accepted that the light travelling from A to M and the light travelling from B to M do so at the same velocity.

From these conclusions, Einstein is able to derive the relativity of simultaneity. His argument, as presented in *Relativity*, is as follows. Imagine one frame of reference K that is fixed relative to the Earth, more specifically that is fixed relative to the ground beside a railway track. Let K* be another frame of reference that is fixed relative to a train travelling at constant velocity in a straight line along the railway track. We will suppose the train to be travelling from left to right. Plainly, then, frame of reference K* will be inertial with respect to K. Einstein asks us to suppose that a single bolt of lightning strikes the rear of the train and the ground directly beneath the rear of the train. We are to regard the event of the lightning striking the rear of the train and the ground directly beneath it as a single event. We refer to this single event as A. At "approximately" the same time, another bolt of lightning strikes both the front of the train, and the ground directly beneath it. Again, we are to regard this as a single event, which we refer to as event B. We are now to imagine an observer O located on the ground, at the point M midway between events A and B. We could verify that O is located at a point mid-way between A and B by laying measuring rods down on the ground from A to O and from B to O. Suppose that the light from A and the light from B arrive at M simultaneously. Then, by Einstein's definition, events A and B are, from the point of view of observer O, simultaneous. But now let us imagine another observer O* located on the train, halfway between the front of the train and its rear. We could, again, verify that O* is located mid-way between A and B by laying down measuring rods along the floor of the train, from A to O*, and from B to O*. Relative to observer O located on the ground, observer O* on the train is moving from left to right. Now, we have already noted that the velocity of light is finite. Therefore, in the time it takes light to travel from the point at which it strikes the ground to observer O located on the ground, the train will have moved some distance relative to the ground. Consequently, observer O* will have moved a little to the right in the time it takes the light from the rear of the train to reach observer O* on the ground. Similarly, because observer O located on the ground is moving to the *left* relative to the observer in the train, the observer on the ground will have (relative to the observer on the train) have moved a little to the left in the time it takes light from the front of the train to reach the observer located on the train. These considerations plainly show that if the light from events A and B reach the observer *on the ground* simultaneously, then the light from the two events will *not* reach the observer *on the train* simultaneously. The light from the front of the train (event B) will reach the observer at the mid-point of the train before the light from the rear (event B). This is an inevitable consequence of the fact that the velocity of light is finite. But Einstein's definition of simultaneity says that events A and B are simultaneous if and only if light from A and B reach a point M midway between A and B simultaneously. So, on Einstein's definition, the events A and B are simultaneous relative to the observer on the ground, but are not simultaneous relative to the observer on the train. More generally, on Einstein's view, there is no such thing as absolute simultaneity, there is only simultaneity with respect to one frame of reference or another.

Let us see how the various premises Einstein has already defended work together to support the relativity of simultaneity. Returning to the example of the train, perhaps the initial common-sense response is that the two events A and B really are simultaneous, and that the point of view of the observer on the ground is correct while that of the observer on the train is wrong. But, if this position is to be maintained, there must be something in virtue of which the observer on the ground is right and the other wrong. It is perhaps initially tempting to say that the observer on the ground is correct because he is at rest whereas the observer on the train is in motion. But this response clearly is not available, since Einstein has already argued that there is no distinction to be drawn between absolute rest and uniform motion in a straight line. It might be suggested that the observer on the ground is correct because they are at rest not with respect to "space itself", but with respect to the events A and B. But a little thought shows us that this response is not available either. The event A, for example lightning striking the rear of the train, is not something that endures for an extended period of time. We can regard it as occurring (more or less) instantaneously. To say that an object is at rest with respect to an *event* means that its distance from the (instantaneous) event remains the same. But what could it mean to say that the distance an object has from an event remains the same even after that event has ceased? It seems it could only mean the distance the object has from the point in space at which the event occurred remains the same. But now we must ask ourselves what it means to say that a particular point in space is the very same point in space over some period of time. It seems the only thing it could mean is that it is at rest with respect to absolute space. But since this is a notion that Einstein rejects, the notion of an observer being at rest with respect to an *event* must also be rejected. So, we have yet to see in virtue of what the observer on the ground could be really correct when he asserts events A and B to have occurred simultaneously.

Einstein said that two events are simultaneous if light reaches a point mid-way between them simultaneously. But it might be suggested that in the time it takes light to travel from the front of the train to observer O, observer O* has moved away from the mid-point between the two events. But then the question arises "Why should we say this about the observer on the train rather than the observer on the ground?" After all, from the point of view of the observer on the train, the observer on the ground is moving steadily to the left. Again, it seems we would only have a reason to say that it was the observer on the train that had moved away from the mid-point if we said that the observer on the ground was at rest with respect to absolute space, and this is a notion that Einstein rejects.

Perhaps the most natural response to make to the case imagined is that it simply shows the untenability of Einstein's *definition of simultaneity*. Einstein suggested that two events are said to be simultaneous if the light from them arrives at a mid-point simultaneously. But we noted that, intuitively, this would only seem to give us a definition of *simultaneity* if the velocity at which light travels in one direction is the same as the velocity with which it travels in the other. Moreover, Einstein himself says that we do not have any *direct* empirical evidence that the light does travel at the same speed in both directions. So if the assumption that the speed of light is the same in both directions leads us to such highly counter-intuitive conclusions as the relativity of simultaneity and the idea that there is no such thing as *the* time at which an event occurred, then it surely very plausible to suggest that the assumption that the velocity of light is the same in both directions ought to be rejected.

But, of course, this is not the move Einstein himself makes. He retains the assumption that the speed of light *is* the same in both directions – despite the fact that there is no direct empirical evidence for it – and therefore accepts the correctness of the relativity of simultaneity. Why? It will be argued that Einstein's decision is in fact explicable on the perspective defended here.

First let us consider the following question: "It is true that intuitively, or from a common-sense point of view, the idea that there must be some objective fact of the matter concerning whether or not A and B occurred at the same time seems almost irresistible, but what empirical evidence is there that this idea is correct?" Presumably, the evidence is of the following sort. Suppose some event occurs, such as a plane crashing into a mountain. We wish to find out at what time this event occurred. Various witnesses to the event all say it occurred at, say, 11.05 pm. We can, for example, suppose that all witnesses to the event say the plane crashed when the clock on the town hall showed 11.05 pm. Because of this inter-subjective agreement, we are perhaps inclined to say that it is an objective fact that the crash occurred at 11.05 pm. But it is plain that empirical evidence of this sort is very limited. All the observers of events such as these, with which we are familiar, have been, as near as does not matter, *stationary* with respect to each other. The velocities of these observers with respect to each other have been, when compared to the velocity of light, virtually zero. Therefore, all such observers have, as near as does not matter, been using the *same* frame of reference. So the empirical evidence we have for our common-sense conviction that simultaneity is absolute is limited: it is restricted to observers sharing (as near as does not matter) the same frame of reference. This empirical evidence provides strong inductive confirmation for the generalization that observers sharing the same frame of reference will agree on simultaneity, but it does not provide confirmation that *all* observers, whether sharing the same frame or not, will agree on simultaneity. It is worth noting that we see here another application of Newton's third "rule of reasoning in philosophy": that inductive inferences are to be applied only to those properties "admitting neither intensification nor remission of degrees". The property of being a frame of reference that is stationary with respect to other frames of reference is a property that does not admit of intensification or remission of degrees, and so, on Newton's third rule, we are entitled to conclude that since simultaneity is the same for frames stationary with respect to each other within the reach of our experiments, this is so for all frames stationary with respect to each other. But since the property of being in motion is a property that admits of intensification or remission of degrees, we are not entitled to conclude from the fact that simultaneity is the same for all the frames that have one specific degree of motion with respect to each other (namely zero motion) that this is going to be true for all frames that have any degree of motion with respect to each other. And, as we have noted, Newton's third rule is explicable within the perspective advocated here.

So, in summary, while we do have extensive empirical support for the claim that observers at rest with respect to each other will agree on simultaneity, we do not have evidence that this is so for observers in frames in motion with respect to each other. So our common-sense conviction in the absoluteness of simultaneity would not seem to be supported by any empirical evidence. On the other hand, while there is no *direct* empirical evidence that the velocity of light is the same in all directions, we saw in the previous section saying it is maximizes independence. So, on the view advocated here, at least, Einstein is right to keep the hypothesis that the velocity of light is the same in both directions, and reject the absoluteness of simultaneity.

EINSTEIN ON THE CONCEPTS OF TIME AND SPACE AS APPLIED TO MOVING BODIES: THE FUNDAMENTAL ROLE OF THE RELATIVITY OF SIMULTANEITY

Einstein's concept empiricism requires him to be able to define expressions such as:

> Object O is d metres long   (1)
> Events A and B were t seconds apart   (2)

in terms of the results of actual or possible observations or measurements. If object O is at rest with respect to the observer, then no special problems arise in specifying a procedure that would tell us its length. We could, for example, simply measure its length with a measuring rod. Similarly, no special problems arise if the two events A and B occur at the same location in space with respect to the observer. We could, for example, simply place a clock at the location at which A occurs, wait until B occurs, and then look at the clock to see how much time has passed.

However, things are not quite so straightforward in verifying (1) if O is in motion with respect to the observer or in verifying (2) if the events do not occur at the same point in space. Let us consider how we might go about determining the difference in time between two events that occurred some distance apart. Suppose, for example, that a bolt of lightning strikes point X and, a short while later, another strikes point Y. The most natural way of determining the time difference between the two events would be for an observer located at X to note the time at which the bolt strikes, for an observer at Y to do the same, and then compare the two times. For example, the observer at X might say the bolt that hit X occurred at 4.00 pm, while the observer at Y said the Y bolt hit at twenty seconds past four. Then we would say the time difference between the two events was twenty seconds. But it is plain that this would only give us an accurate measure of the time between the two events if the clock at X and the clock at Y kept *the same time*. But now, what does it mean to say that the two clocks keep the same time? It plainly means nothing other than that the two clocks strike 4.00 pm, 4.01 pm and so on *at the same time*. Hence, the claim that they keep the same time is dependent on a claim of simultaneity. But simultaneity is, for Einstein, relative to frame of reference. So the time difference between two events will also be dependent on the frame of reference from which the two events are observed. These considerations show, Einstein asserts, that we are not entitled to automatically assume that there is some frame independent fact of the matter concerning the time interval between the two events.

Similar remarks apply to the measurement of the length of an object that is in motion with respect to an observer. How might we go about establishing the length of an object at motion with respect to an observer? Einstein suggests the following method. Suppose, to make things specific, that an observer located on the ground is confronted with the problem of determining the length of a train moving with uniform motion in a straight line. The observer located on the ground is able to measure things that are at rest with respect to the ground by a measuring rod. So, the problem of measuring the length of the train reduces to the problem of ascertaining that the train has the same length as some object, such as a railway platform, stationary with respect to the ground. Einstein suggests that we are entitled to assert that the length of the train is the same length as the platform if the rear of the train reaches one end of the platform and the front end of the train the other end of the platform *simultaneously*. So we see that judgements of the length of moving bodies are dependent on judgements of simultaneity. But simultaneity is relative to frames of reference. So, we cannot automatically assume that measurements of length made from different frames of reference will necessarily give the same result.

We can summarize Einstein's suggestions concerning the measurement of time and space intervals as follows:

> Let clock 1 be located at the point in space and time at which event   (T)
> A occurs and let clock 2 be located at the point in space and time
> at which event B occurs, and let clock 1 register the time at which
> A occurs and clock 2 register the time at which B occurs. Then the
> difference in the time at which events A and B occurred is *t* seconds
> if and only if the difference in the times registered by the two clocks
> is *t* seconds and the two clocks are synchronous, that is the two
> clocks strike 4.00 pm simultaneously, strike 4.01 simultaneously
> and so on.
>
> Let R be a rigid rod that is in motion with respect to frame of   (L)
> reference K and Ra and Rb be the end points of R. Let R* be a rigid
> rod that is stationary with respect to K and R*a and R*b be the end
> points of R*. Suppose R and R* to both be moving with uniform
> velocity along the same straight line. Let A be the event of Ra and
> R*a occupying the same position in space and let B be the event
> of Rb and R*b occupying the same position in space. Then rods R
> and R* have the same length if and only if the events A and B occur
> simultaneously.

(T) and (L) make it clear how the ways of determining time intervals, and length, favoured by Einstein depend upon the notion of simultaneity. It is worth briefly discussing the *status* of the claims (T) and (L). Einstein writes as if these claims were analytic definitions of our notions of length and time; that they gave us the content of our notions of length and time and that, unless length and time were spelled out in terms of the results of observable results of measurements, our notions of length and time would be "meaningless". But we do not have to accept these strongly concept-empiricist ideas to accept the soundness of Einstein's overall argument. In order for Einstein's argument to be sound, it is sufficient that (T) and (L) be *epistemically necessary* truths. And I think we can agree that they are epistemically necessary without subscribing to Einstein's concept empiricism. Their truth seems to follow merely from the meanings we attach to the terms "time", "space" and "simultaneity".[3] Since no *empirical* claim is made in (T) and (L), it is not necessary to show that their acceptance can be explained on the account advocated here.

THE REMOVAL OF THE CONTRADICTION MOTIVATING EINSTEIN, TIME DILATION AND LENGTH CONTRACTION

Earlier we noted that a contradiction arose between the following two principles:

> *The principle of relativity*: For any frames of reference K and K*, if, relative to K, K* is inertial and non-rotating, natural phenomena run their course with respect to K* according to the same natural laws (including the law of the constancy of the velocity of light) as they do with respect to K.

> *The classical principle of the composition of velocities*: Let K* be a frame of reference inertial with respect to K and moving in direction x at velocity v with respect to K. Then, if O is an object moving in direction x at velocity u with respect to frame K*, then O will be moving in direction x at velocity W $= u + v$ with respect to frame K. (Conversely, if O is moving in direction x at velocity u with respect to K, it will be moving at velocity $W^* = u - v$ with respect to frame K*.)

Since these two principles contradict each other, they cannot both be maintained. Einstein says it is the classical principle of the composition of velocities that is to be rejected. And he says that, in the light of (T) and (L) and the relativity of simultaneity, we no longer have any reason to believe that the classical principle is true.

It is worth considering in a little detail just why this should be so. Let us consider what would need to be the case for the classical principle to be correct; more specifically, for it to be the case that if O is moving at u with respect to K*, then it must be moving at $u + v$ with respect to K. As is the case for all statements of all sorts, Einstein holds that any statement of the form "object O is travelling at velocity v" must be capable of being given an analysis in terms of the results of actual or possible measurements. Although such an analysis has not yet been explicitly given, it follows directly from the statements (T) and (L) given above. Velocity is given by the formula Vel = $\Delta l/\Delta t$, where Δl is the distance travelled by a given object in time Δt. We can therefore use (T) and (L) to construct a procedure for determining the velocity of an object in motion with respect to a given frame of reference. But we have already noted that since simultaneity is relative to frames of reference, there is no guarantee that determining the length of time intervals from different frames of reference will necessarily give the same result. Therefore, there is also no guarantee that determining velocity from different frames of reference will necessarily give the same result. More specifically, let us assume, using Einstein's own example, that K is stationary relative to the Earth and K^* is stationary relative to a train moving at uniform velocity in a straight line. We imagine a person walking along inside the train, in the same direction as the motion of the train. Our task is to calculate the velocity of the walker relative to the train. Since velocity is given by the formula Vel = $\Delta l/\Delta t$, we can calculate the velocity of the walker by calculating the time Δt it takes the walker to travel a given distance Δl, for example the length of a single carriage. But since it is a consequence of (T) and (L), and the doctrine of the relativity of simultaneity, that measurements of time and distance are relative to frames of reference, the measurements of the time it takes a walker to get from one end of the carriage to the other, and of the length of the carriage, that would be obtained by an observer on the train cannot be assumed to be the same as the corresponding measurements that would be obtained by an observer on the ground. Hence, it cannot be assumed that the observers will necessarily obtain the same measurement of the velocity of the walker relative to the train.

It is easy to see that these considerations show that the classical principle of the composition of velocities can no longer be assumed to be correct. The classical principle asserts that if frame K^* is moving at v with respect to frame K, and object O is moving at u with respect to K^*, then O will moving at $u + v$ with respect to K. Now, if the train is moving at v with respect to K, then for it to be the case that an observer in K finds O to be moving at $u + v$, it must be the case that the observer in K finds O to be moving at u with respect to the train. Now, we know that an observer located on the train finds the walker to be moving at u. Therefore, if the classical principle is to be correct, the observer on the ground must find the walker to be moving at the same velocity with respect to the train as does the observer on the train. But it is just this assumption that we are not entitled to make, given (T), (L) and the relativity of simultaneity. Therefore, if we accept (T), (L) and the relativity of simultaneity, we cannot automatically accept the classical principle of the composition of velocities. Hence, the derivation of the contradiction fails to go through, and the problem with which Einstein was concerned has been dissolved.

The following question therefore naturally arises: "Suppose it to be the case that, relative to K^* – that is relative to the train – the walker is moving forward at velocity u, how fast will an observer located on the ground find the walker to be moving relative to the train?" We determine velocity according to the formula Vel = $\Delta l/\Delta t$. Therefore, the question "How fast is the walker moving?" gives rise to the following two logically prior questions: "If an observer on the train measures a given object O to have a particular length Δl, how long will that object be found to

157

be by an observer on the ground?" and "If an observer on the train finds the time interval between two events to be Δt, how long will an observer on the ground find the time interval to be between the same two events?" It is possible to answer these two questions by deriving the answer that would be obtained by an observer on the ground using (T), (L), Einstein's definition of simultaneity and the assumption that the velocity of light is the same in both directions. As the derivation of the values is essentially mathematical, and as we are here concerned only with *ampliative* inferences, we need not here concern ourselves with the details of the derivation.[4] But the derivation gives the following answers. If a given metre rod is stationary with respect to frame K*, and frame K* is moving at velocity v with respect to frame K, then the length l of the metre rod in K*, as measured by an observer in K, is given by the formula:

$$l = l \times (1 - (v^2/c^2))^{\frac{1}{2}} \tag{LC}$$

where c is the velocity of light. The formula (LC) tells how the length of an object in motion relative to an observer will be found to contract. If the time between two events that occur at the same point in space relative to frame K* (such as two successive ticks of a clock that is stationary relative to frame K*) is one second when measured from within frame K*, then the time t between the two ticks, when measured from frame K, is given by:

$$t = (1 - (v^2/c^2))^{-\frac{1}{2}} \tag{TD}$$

The formula (TD) tells us by how much the time in a frame of reference in motion relative to an observer will be found to slow, or dilate.

There is one more formula with which we will here be concerned. The formula (TD) tells us how much more slowly a clock moving at velocity v will be found to be running as compared to a clock that is stationary with respect to an observer. It is possible to derive from (TD) a formula giving by how much such a clock will be retarded after it has travelled a distance d. This is as follows:

$$R = d/v \, [1 - (1 - (v^2/c^2))^{-\frac{1}{2}}] \tag{Ret}$$

where R is the extent of the retardation of the moved clock after it has travelled distance d at velocity v.

THE EMPIRICAL SUCCESSES OF THE SPECIAL THEORY OF RELATIVITY

The equations (TD) and (Ret) both lead to testable empirical predictions that have received observational confirmation.

(TD) asserts that if an object is moving relative to an observer, then, from the point of view of the observer, time will appear to be running more slowly for the moving object. This leads to the prediction that, for the moving object, all *physical processes* will appear to be running more slowly. Normally, of course, any such slowing of the physical processes taking place within an object is too slight to be observed, since most moving objects in our immediate surroundings are moving at velocities extremely low compared to that of light. However, in particle accelerators it is possible to accelerate some particles to close to the speed of light. There are some such particles, *muons*, which normally have extremely brief life-spans. The formula (DT) leads to the testable prediction that if muons are accelerated close to the speed of light, their life-spans will be detectably increased. And this has been observed to be the case (Frisch & Smith 1963: 342–55).

The formula (Ret) asserts that if a clock is moved over some distance d, it will "lose time" as compared to another clock that is not moved. This leads to the prediction that if two clocks are synchronized and one moved and then returned to the other, the moved clock will be found to be running behind the other. It turns out that if two extremely accurate caesium clocks are synchronized, and one flown around the Earth in a jet, this retardation of the moved clock ought to be detectable. Again, experiments have confirmed this predicted effect (Hafele & Keating 1972).

Both these predicted effects are surprising, and would appear to be *a priori* highly improbable. Therefore, an explanation of these predictive successes of special relativity is required.

HOW ARE THE EMPIRICAL SUCCESSES OF SPECIAL RELATIVITY TO BE EXPLAINED?

On the view advocated here, we are to explain the success of empirical hypotheses by showing them to be independent of the data, or to have intra-DEC independence, or to be warranted by the AIM inference. In this chapter so far it has been argued that the various ampliative, non-mathematical steps in Einstein's argument can indeed be interpreted in this way. It appears, therefore, that the predictive success of the special theory is explainable on the view advocated here.

There is, however, one controversial aspect of our interpretation of Einstein's argument that needs further examination. This aspect concerns the thesis of the constancy of the velocity of light. We need here to distinguish between the following two claims:

The average, two-way velocity of light takes the same value for all (i) observers, in all inertial frames of reference. This common value is conventionally denoted by c.

The velocity of light is the same in all directions, in particular, its (ii) velocity from A to B is the same as that from B to A.

Point (i) is supported by the results of the Michelson–Morley experiment and the principle of relativity. There is, however, no *direct* experimental evidence for the correctness of (ii). Hence, a claim that appears to play a crucial role in the derivation of the formulae of the special theory is not supported by any direct experimental evidence.

Earlier in this chapter it was argued that although (ii) is not supported by any direct empirical evidence, it nonetheless counted as a "good" hypothesis on the perspective advocated here. In particular, it was argued that (ii) is more independent of the data than any alternative hypothesis. So, on the view given here, a theory containing (ii) is, all other things being equal, more likely to lead us to correct observations than its rivals. This is able to provide us with a probabilistic explanation of why it is that the observable predictions derived from Einstein's special theory should turn out to be correct.

We should recall, however, that Einstein himself did not regard the hypothesis that the velocity of light was the same in both directions as a hypothesis that was literally true. He explicitly asserted that it is not a "*supposition nor a hypothesis* about the physical nature of light" at all, but a "*stipulation*" (Einstein 1920: 23). Subsequent authors have interpreted this to mean that for Einstein it is merely a *convention* that the one-way velocity of light is the same as its average two-way velocity.

The fact that the one-way velocity of light cannot be empirically verified has given rise to a considerable literature. This literature has explored questions such as "Is the claim that the one-way velocity of light is equal to c merely a convention, or is it something that is capable of being true or false?" and "Can we produce a version of special relativity that does not say anything about the one-way velocity of light?"

However, from our point of view, the issue of the empirical *verifiability* or otherwise of the one-way speed of light does not matter. Our concern is to explain certain novel predictive successes of science: in the present case, the successful prediction of phenomena such as time dilation and clock retardation. The strategy adopted is to show that the special theory of relativity has a high degree of independence from the data, and hence that it is highly unlikely that it should merely be due to chance that the data conform to this theory. This makes it likely that there exists a propensity for the data to conform to the theory. But the crucial point to note here is that to say the data has a propensity to conform to the theory is not to say anything at all about the truth-value of the sentences in the theory. Neither is it to make any claim about their independent verifiability or testability. To say a theory is highly independent is merely to say, roughly, that it states the simplest pattern to be discerned in the data and hence (by the argument of Chapter 4) the pattern most likely to be continued. On the face of it, there is no reason why each sentence in such a theory would need to be independently testable for it to be the case that the theory as a whole identifies the simplest pattern to be discerned in the data. So, on the view advocated here, the issue of the independent testability or otherwise of the one-way speed of light is irrelevant.

CONCLUDING REMARKS

In this chapter it has been argued that we can explain the surprising novel successes of the special theory of relativity using the conceptual apparatus developed earlier in the book. More specifically, it has been argued that the notion of the independence of theory from data, the notion of simple DECs and the AIM inference can explain the phenomena of time dilation and clock retardation. This strengthens the claim that the apparatus developed here can explain the phenomena with which we are concerned.

CHAPTER 8

Mendelian genetics

In the previous two chapters, we were concerned with applying the ideas developed in the bulk of this book to cases from the history of science. In Chapter 6 we considered Newton's arguments for his laws of motion and his theory of gravitation. In Chapter 7 we examined Einstein's arguments for special relativity. Of course, both these chapters took as their subject matter examples from physics. It is therefore natural to wonder whether the point of view developed here applies to sciences other than physics. Perhaps the most obvious non-physical science to look to is chemistry. I have elsewhere argued that the notion of the independence of theory from data can account for the transition from the phlogiston theory of Cavendish to the oxygen theory of Lavoisier (Wright 1991). Consequently, rather than consider here another example from chemistry, we consider a case from the biological sciences, specifically Mendel's development of genetics.

THE PHENOMENA MENDEL WAS CONCERNED TO EXPLAIN

We may identify four phenomena that were available to Mendel when he developed his theories. These were:

> *Phenomenon A The existence of "pure breeds".*[1] It was known in Mendel's day that not all plants or animals that *looked* the same would necessarily have the same type of offspring. We can imagine two classes of plants P1 and P2 that all looked exactly the same, both in the sense that every member of P1 looked exactly the same as every other member of P1 and every member of P2 looked exactly the same as every other member of P2 and also in the sense that every member of P1 looked exactly like every member of P2. But Mendel knew that, despite the fact that every member of the two classes looked exactly like every other member, the two classes may still produce very different offspring. Suppose new generations of plants are produced by fertilizing members of P1 only with other members of P1 and fertilizing members of P2 only with other members of P2. Then, if P1 is a pure breed, all the offspring of P1 will look exactly like their parents, that is exactly like the members of P1, as will all subsequent generations produced only from P1

and its offspring. But if P2 is not a pure breed, then offspring produced from within P2 need not all look like the members of P2. The members of P1 will "breed true" while the members of P2 will not. A class of organisms, whether plants or animals, constitute a "pure breed" if and only if they "breed true". So to assert that some class of organisms is a pure breed is not just to make a claim about the appearance of its members, it is also, at least, to make a claim about how its descendants would look.

Phenomenon B *The fact that when different pure breeds are crossed, all of the resulting hybrids possess only the characteristic of one of their pure-breed parents.* There can be, within a single species, a number of different pure breeds. Consider, for example, the species of garden pea studied by Mendel. He noted that one pure breed of this species of pea produced only peas with smooth skins, another only peas with wrinkled or "angular" skins (Mendel 1913: 339–40).[2] When a collection consisting entirely of a pure breed of smooth-skinned peas was crossed with a collection entirely of a pure breed of peas with wrinkled skins, the resulting hybrid peas were always found to *only* have smooth skins. Similarly, one pure breed of the peas had pods with green skins, another had pods with green skins. When these pure breeds were crossed, the resulting hybrid peas always had only pods with yellow skins. (It is conventional to denote the first generation of any organism obtained by crossing two breeds by the symbol "F_1".)

Phenomenon C *When members of an F_1 generation obtained by crossing two pure breeds are themselves crossed with other members of that same F_1 generation, some, but not all, of the resulting offspring (the F_2 generation) exhibit the characteristic that was "lost" in the F_1 generation* (Mendel 1913: 344–7). As we noted above, Mendel obtained a hybrid generation of peas all of which had smooth skins by crossing one pure breed of smooth-skinned peas with another pure breed of peas with wrinkled skins. The result was a generation of peas all of which had smooth skins. But when the members of this hybrid generation were crossed with other members of that same generation, some, but not all, of the peas in the next (F_2) generation had wrinkled skins. The characteristic of wrinkled skin, which had been temporarily lost in the F_1 generation, reappeared in some of the members of the next generation.

Phenomenon D *The exact numbers, as recorded by Mendel, of the peas in the F_2 generation that possessed the characteristic that had been temporarily lost in the F_1 generation* (Mendel 1913: 344). When Mendel had collected the peas from the F_2 generation, he noted that of a total of 7,324 peas, 5,474 had smooth skins while 1,850 had angular, wrinkled skins. That is, 74.74 per cent of the peas had smooth skins while 25.26 per cent had wrinkled skins, or the ratio of peas with wrinkled skins to those with smooth skins was $1:2.959$. Mendel also performed a similar experiment with pod colour. He collected a total of 8,023 pods from his F_2 generation. Of these, 6,022 were yellow while 2,001 were green. That is, 24.94 per cent of the pods were green while 75.06 per cent were yellow. The ratio of green pods to yellow pods was $1:3.01$.

A RECONSTRUCTION OF MENDEL'S INFERENCES
FROM THE PHENOMENA

In this section a reconstruction of Mendel's inferences from these four phenomena will be given.

First, a prefacing remark. Unlike, for example, Newton or Einstein, who gave us detailed accounts of the inferential steps involved in arriving at their theories, Mendel seems (at least to the present author) to have given us only a rather incomplete account of his inferential steps. Consequently, our discussion of Mendel will be more of a "rational reconstruction" of his inferences than was the case with Newton or Einstein.[3]

A fair number of the key ideas of Mendel's theory are plausibly implicit in the very notion of a "pure breed". It is a part of the notion of a pure breed that it is possible to have two collections A and B of organisms that are observationally indistinguishable, but which are such that one of them is a pure breed, while the other is not. To say that something is a pure breed is therefore to do more than merely assert that it has certain observational properties. It is at least to say that all of its *descendants* would share its observable properties. But, very plausibly, it is also to do more than that. Presumably, it is not a part of the common-sense notion of the difference between one collection of organisms that is a pure breed and another observationally indistinguishable group that is not that one group will *just happen* to have all its descendants the same while the other group will *just happen* to show variation among its descendants. We presumably believe there is something about the pure breed which ensures or causes it to be the case that all its descendants will be the same, and we also believe that there is something about the collection that is not a pure breed that brings it about that some of its descendants will not be like their parents. So, for example, if A is a collection of pea plants all of which have yellow pods, and we assert that A is *not* a pure breed, then we presumably believe that there exists inside the members of A some factor that will bring it about that some of the descendants of A will have green pods. And we presumably believe this to be so even though there is nothing observable about the members of A indicating the presence of such a factor. So, plausibly, to assert that some collection of organisms is or is not a pure breed is to assert the absence or presence of some not-currently-observable factor that will be responsible for aspects of the appearance of the offspring of the members of the collection.

There is another (very obvious) point to make about the ideas implicit in the notion of a "pure breed". Suppose again that A is some collection of organisms that is *not* a pure breed. It might be, for example, a collection of pea plants all of which have yellow pods. There will, nonetheless, be present in the members of that collection some factor that is capable of producing *green* pods. And this, we feel, is *why* some of the offspring of the members of A do turn out to have green pods. Obviously implicit in this idea is the belief that the offspring of the members of A *inherit* the factor producing green pods from their parents, that is that the members of A *pass on* to their offspring some unobservable "something" that will cause some of them to have green pods.

Finally, one more point that is plausibly implicit in the concept of a pure breed. Consider again a collection of organisms that is not a pure breed, such as a collection of peas with only yellow pods. We have noted that it just seems to be "common sense" that there must be present in the members of that population some *unobservable* "something" that has the power to produce green pods. This unobservable "something" resides within (at least some of) the members of the collection, and is capable of being passed on to its offspring. And it will cause at least some of those offspring to have green pods. And so we may say that within any given pea plant – presumably within any organism of any kind – there are two "realms" or "levels": one level or realm is of the observable characteristics of the plant itself (its pods, leaves, flowers, etc.), but there is also another, unobservable realm, the elements of which are capable of producing the components seen at the observable level and which are also capable of being passed on from parent to offspring.

So, in summary, it seems that plausibly implicit in the very idea of a "pure breed" are the following ideas:

1. Whether or not the members of some class constitute a pure breed or not is not determined by their observable characteristics. At the very least it is determined by the nature of the offspring they will have or would have had. But it also seems to be common sense that there must also reside in organisms some non-observable "something" that will be responsible for the characteristics of the offspring of those organisms.
2. Organisms pass on these unobservable "somethings" to their offspring.
3. We therefore may distinguish two "levels" or "realms" within organisms: in addition to the observable realm, there is also some unobservable realm, the elements of which are passed on to offspring, and which confer upon those offspring at least some of their observable characteristics.

The beliefs 1–3 were certainly accepted by Mendel, but they also seem to be just common-sense inferences from the very notion of a pure breed. Here it will be argued that the inferences from the phenomena implicit in the notion of a pure breed to beliefs 1–3, can be explained on the account given here.

We are confronted with the following empirical fact. We can have two classes of organisms, A and B, such that all members of the two classes resemble every other member of the two classes; yet when the members of class A are cross-fertilized with other members of class A, the offspring always have the same observable characteristics as their parents, but when the members of class B are cross-fertilized with other members of class B, the offspring are sometimes different from their parents. How are we to respond to this observed fact? There seem to be, broadly speaking, two possibilities:

Either: there is no "explanation" of this fact – this is just the way things will happen.

Or: there is some explanation.

Let us consider the consequences of saying that there is no explanation. Remember that the members of A are observationally indistinguishable from the members of B. The first generation of descendants of A will all resemble the members of A, as will all the members of the second generation, all the members of the third generation and so on. But some members of the first generation of organisms descended from B will have observable characteristics that are not among the observable characteristics of B, as will some members of the second generation, the third generation and so on. The number of generations descended from B that will possess observable characteristics not to be found in the original population of B itself is plainly very large, and is plausibly potentially infinite. Similarly, the number of generations descended from A, all of which resemble A in all observable respects is also, plausibly, potentially infinite. Therefore, if the characteristics of all the generations descended from A and B were simply all accepted as unexplained facts, we would achieve no degree of independence from the data at all: we would simply have a very large – possibly potentially infinite – collection of data. If we are to achieve any degree of independence from the data, we must give some kind of explanation of the characteristics of the subsequent generations.

Explanations of the characteristics of subsequent generations are of two sorts: those that say that the subsequent generations inherit their characteristics from their parents, and those that offer some alternative account. Let us begin by considering the "alternative" explanations.

The alternative explanations allow that the characteristics of the subsequent generations do indeed have causes, but they do not assert that the causes lie in the original stocks of A and B. Where, then, do the causes lie? Let us consider the appearance in generations descended from B of characteristics not present in the original stock of B. It is always possible to explain the appearance of some characteristic, such as green pods, by saying that it was caused by something in the environment of the members of the descendants of B – perhaps something in the air or the water or the soil. But there is still something that any such approach would fail to explain: why is it that the descendants of *some* collections of organisms seem never to have different observable characteristics from their ancestors? If the factor producing a different appearance (such as green pods) was, for example, in the water or air or soil, it would seem that exposure to the same water, air or soil as those strains that did have green pods ought to at least sometimes produce green pods in the pure strains. Therefore, saying that the factor producing the green pods did *not* reside in the original collection of parent organisms, but in the external environment, leaves it as an unexplained fact that there are some collections of organisms the descendants of which never display novel features. On the other hand, if it is asserted that organisms only get their characteristics by inheriting them from their parents, this fact becomes explainable. On this view, the reason why the descendants of the pure breed never display green pods is because the factor producing the green pods was not present in the original parent stock. The reason why the descendants of the non-pure breed do sometimes display novel features such as green pods is because the factor producing the green pods was present in at least some of the members of the original parents.

Of course, the above explanation leaves it unexplained why it is that some collections of original parent stock do posses a factor for producing, say, green pods while others do not. But it seems intuitively clear that the alternative explanation – that the factor producing the novel characteristics lies in the environment outside the plants – leaves *more* unexplained. By saying that the factor responsible for the novel characteristics lies inside the organisms, it is only necessary to postulate one unexplained fact in order to explain why it is that some collections continue to "breed true" throughout all subsequent generations, which is that the factor producing the novel characteristic was not present in the original parent stock. But in order to explain it on the alternative view, it is necessary to postulate a *series* of unexplained facts: that the factor capable of producing the novel characteristic did not enter the organisms from the environment in the first generation, in the second generation, in the third generation and so on. Therefore, the number of unexplained facts that must be accepted in order to explain the same phenomena – the observable characteristics of descendants – is vastly greater if it is asserted that the factor responsible for those characteristics lies outside the parent organisms. That is, the *ratio* of facts accepted to facts explained is much greater. Therefore, the degree of independence of theory from data is much greater for the hypothesis that the factor lies within the parents than it is for the hypothesis that it does not. So the preferability of the former hypothesis is explainable within the point of view advocated here.

There is one more belief, plausibly implicit in the notion of a pure breed, that needs to be accounted for on the view advocated here, and that is the belief that there is an unobservable realm or level of entities present in organisms that are responsible for the observable characteristics of those organisms. That there are *some* such unobservable entities or factors seems to be made plausible by the following argument. Consider two populations A and B. Suppose the members of A and B to have all the same observable properties, but while the descendants of A all have all the observable characteristics of the members of A, some of the descendants of B have characteristics not found in any members of B. It has been argued above that this shows it to be more likely that there is present in the members of B some factor responsible for these characteristics, which they pass on to their offspring. So, let us accept that there is some factor present in the members of B that is not present in the members of A, and call this factor G. At least some of the members of B have G, but none of the members of A have G. But the members of A and B have all the same *observable* properties. Therefore, factor G must be unobservable. Hence, at least some of the factors passed on from parents to offspring are unobservable entities, or belong to some unobservable realm or level. This argument would appear to be a purely deductive one: it follows from the fact that if the difference between A and B in their powers of transmitting characteristics to their offspring is not observable, it must be unobservable. Since this would seem to be a purely deductively valid inference, it is not necessary for it to be explainable on the view advocated here.

So, in summary, it has been argued that, using the inference to theories that maximize independence and a deductively valid inference, it is possible to derive from the familiar data implicit in the notion of a "pure breed" the ideas that there are unobservable factors or entities that reside in at least some organisms, that these organisms pass on these factors or entities to their offspring, and that these unobservable entities are responsible for at least some of the observable characteristics of those offspring.

Let us now consider phenomenon B mentioned above (page 168). This phenomenon involved the crossing of two different pure breeds within a single species. Mendel found that when, for example, one pure breed of peas with green pods was crossed with another pure breed that had only yellow pods, the first generation of descendants all had yellow pods. However, he also found that when members of the first generation were crossed with each other, there were some members of the second generation that produced green pods. By the reasoning given in the previous paragraphs, it was therefore rational to conclude that those members of the second generation that had green pods inherited a factor producing the green pods from their parents, and there was, therefore, present in at least some members of the first generation an unobservable "something" that was a factor capable of producing green pods. We will continue to refer to this factor as G. Let us refer to those members of the *first* generation of descendants that possessed this unobservable factor G as Gen1(G). Now, plainly, since *all* members of the first generation of descendants had yellow pods, a factor producing yellow pods must also have been present in all members of that generation. We will call this factor Y. Hence, plainly, the members of Gen1(G), since they are a subset of that generation, must also have within them factor Y. That is, the members of Gen1(G) must have within them *both* factors G *and* Y. But *all* the members of the first generation, and so all the members of Gen1(G), have only yellow pods. Mendel draws from this the conclusion that *whenever* factors Y and G are both present in a single organism, the resulting organism will display only the characteristic associated with Y, that is yellow pods. This inference is plainly a straightforward induction. It was not precisely known just how many members of Gen1(G) there are. But since 2,001 members of the second generation of descendants had green pods, there must at least have been 2,001 members of the first generation with factor G as well as factor Y. Yet all of them had yellow pods. It is highly unlikely that this could have been merely due to chance. So, the inductive generalization that whenever the factors G and Y occur in the same organism it will have yellow pods is, on the account advocated here, more likely to lead to correct observations than other generalizations.

Mendel referred to the "Y" factor as "dominant" and the "G" factor as "recessive" (Mendel 1913: 342). Of course, to describe one characteristic as "dominant" (or another as "recessive") is not intended as an *explanation* of its behaviour; it is merely to state that, as a matter of fact, when the two factors are present in an organism, one of them will be manifested in the observable features of the organism, while the other will not.

Mendel found that the other characteristics of the peas were also either dominant or recessive (Mendel 1913: 343). For example, he found that when one pure strain of peas with smooth skins was crossed with another with angular skins, the first generation of descendants all had smooth skins, while the characteristic of angular skins showed up again in some members of the second generation. So the characteristic of smooth skins was dominant, that of angular skins recessive. This result plainly provides inductive support for the more general conclusion that whenever there are two, mutually incompatible, ways in which an organism can be (smooth skinned/angular skinned, yellow/green, short/tall and so on) one of the members of the pair will be dominant, the other recessive. Since this more general conclusion is, again, an application of induction, its acceptance is explicable on the account advocated here.

It is appropriate at this point to introduce the notation Mendel himself devised for representing these ideas.[4] Mendel denoted factors determining dominant characteristics by capital letters, and those determining recessive characteristics by lower-case letters. The factors for any pair of incompatible ways in which an organism could be, such as tall/short, yellow/green, smooth-skinned/angular-skinned and so on, were denoted by the same letter. So, for example, if the factor for having yellow pods was denoted by "A", the factor for having green pods was denoted by "a". If the factor for having smooth skins was denoted by "B", the factor for angular skins was denoted by "b", and so on. We have already noted that Mendel believed that the members of the class we are calling Gen1(G) contained the factors *both* for yellow pods and for green pods. We will, following Mendel, from now on refer to these two factors as "A" and "a" respectively. Mendel represented the factors they possessed by the letters "Aa".

As we observed previously, Mendel found that although when a pure breed of yellow-podded peas were crossed with a pure breed of green-podded peas, the *first* generation of descendants only had yellow pods, in the second generation of descendants the green pods reappeared. Mendel also recorded the numbers of green-podded peas in the second generation. Of the 8,023 peas he counted, 6,022, or 75.06 per cent, had yellow pods, while the remaining 2,001 peas, or 24.94 per cent, had green pods. The ratio of green pods to yellow pods was 1 : 3.01. Similar results were obtained when he counted the number of smooth-skinned peas and rough-skinned peas in the second generation: of the 7,324 peas he counted, 5,747, or 74.74 per cent, had smooth skins, while the remaining 1,850, or 25.26 per cent, had angular skins. The ratio of angular-skinned peas to those with smooth skins was 1 : 2.959.

Perhaps the first thing that strikes us about these figures is that the ratios are both very close to 1 : 3. And, of course, Mendel concluded that the correct or real ratio was indeed 1 : 3, and that the slight deviation from this ratio was due to error or randomness of some kind.[5] Plainly, this can be regarded as showing a preference for low whole numbers. Plausibly, both the arguments for the preferability of low whole numbers discussed in Chapter 5 would have been available to Mendel. Let us assume that the hypothesis that the real ratio is 1 : 3 is *permissible*, given the possibility of error due to interfering factors or "noise". Of course, there will also be other hypotheses that are permissible, such as that the real ratio is 1 : 3.002 or 1 : 3.054 and so on. But the *a priori* probability that the data should be describable by a ratio mentioning just a single low whole number (three) is very much lower than the probability that it should be describable by a ratio mentioning a number with two or three or four decimal places. It is also, of course, possible that the fact the ratio was so close to 1 : 3 had already suggested to Mendel the likely mechanisms of inheritance. And it is a consequence of the theory of those mechanisms that the true value of the ratio should be 1 : 3 *precisely*. So it was not likely to be due to chance that the ratio should lie so close to a low whole number, and it is a consequence of why it does lie so close that it should be equal to the low whole number precisely. Consequently, by the argument of Chapter 5 (pages 105–10), the hypothesis that it is the low whole number precisely is most likely to lead to correct predictions.

We have already argued, using the inferential processes advocated here, that the members of the class we have denoted by Gen1(G) have within them *both* the factor for producing yellow pods and the factor for producing green pods. Using Mendel's own system of symbolic representation, the factors they contain within them are represented as "Aa", although only the dominant factor "A" – the factor producing yellow pods – is manifested in the observable characteristics of the plants in Gen1(G). But the point we need to note at the present is that the members of Gen1(G) have *two* factors within them: one unit of the "A" factor and another unit of the "a" factor. This raises the following question: "Are the members of Gen1(G) alone in carrying within themselves two factors, or do some, or even all, other organisms carry within them two such factors?"

At this point it is important to clarify just what is in question. The question is not whether all peas contain within them two different *types* of factor, such as "A" and "a", but whether they contain within them two *tokens* or instances of a factor, whether those two tokens or instances are of the same type or different types. So, for example, the question is whether all plants contain within them some pair of factor-tokens, such as "AA", "Aa" or "aa", or whether some contain only a single factor-token, such as "A" or "a".

In considering this question, let us suppose that there is some pea plant O that contains only one factor-token. We will, for simplicity, assume it is the factor "A". We will also assume this particular plant is a member of the first generation of descendants from the crossing of the two initial stocks of pure breeds; that is, it is a member of the same generation as the class Gen1(G). Now, we have already noted that the members of Gen1(G) have within them two factor-tokens: "A" and "a". But the pea plant O is a product of the same causal process – the crossing of the members of the two initial stocks of pure breeds – that gave rise to the members of Gen1(G). So it would be expected that the same type of causal process would give rise to the same type of outcome. Therefore, if it were asserted that O possessed a different number of factor-tokens from those possessed by the members of Gen 1(G), this would require the postulation of a distinct causal process from the one that confers on organisms two tokens. Hence, the hypothesis that O has only one factor-token requires us to postulate more causal processes than would the hypothesis that it has two. Consequently, the hypothesis that O has only one factor-token leads to an explanatory theory that is less independent of the data, and so, on the account offered here, less likely to lead to correct predictions, than the hypothesis that it has two.

This argument leads us to prefer the hypothesis that all members of the second generation have two factor-tokens. But, of course, we are not entitled to draw from this the conclusion that *all* organisms have two factor-tokens: after all, the parents of the members of the first generation are different. One parent has green pods, the other yellow. Perhaps offspring only have two factor-tokens when they are the offspring of different parents; perhaps the offspring of identical parents only have one factor-token. But it will be argued below that this involves more complicated causal processes than asserting that organisms always inherit two factor-tokens, one from each parent.

Let us return to the two original pure-breed stocks, one of which has green pods, the other of which has yellow pods. But pure breeds *breed true*, that is all organisms descended *solely* from the members of a pure breed possess only the observable characteristics of that pure breed. So all descendants of a pure breed of green-podded peas, provided they were descended *only* from that pure breed and did not have any parents from outside that pure-breed stock, would have only green pods. This behaviour makes the hypothesis that a pure breed of, say, green-podded peas contains within it a factor for yellow pods explanatorily redundant. Therefore, on the view advocated here, the hypothesis that a pure breed of, say, green-podded peas contains somewhere within it factors for yellow pods is less independent of the data than the hypothesis that it contains only factors for green pods. By parity of reasoning, the most independent hypothesis is that a pure breed of yellow-podded peas does not contain within it a factor for green pods. That is, the hypothesis to be preferred, on the view advocated here, is that the pure breed of green-podded peas contains only factors for green pods, and that a pure breed of yellow-podded peas contains only factors for yellow-podded peas. But it has also been argued that every member of the first generation of plants obtained by crossing these two pure breeds contained factor-tokens for *both* yellow pods and green pods. Plainly, the only available explanation of this is that each member of that generation inherited a factor-token from each of their parents: they inherited one from their mother, another from their father. So, we can at least say that if the parents belong to two different pure breeds, the offspring will inherit a factor-token from each parent. But what about if the parents do not belong to different pure breeds? Here we may distinguish two possible hypotheses:

> *Hypothesis 1*: If the parents belong to two different pure breeds, each parent contributes one factor-token to the offspring (the offspring ends up with two factor-tokens); if the parents do not belong to different pure breeds, only one parent contributes a factor-token to the offspring (so the offspring ends up with only one factor-token).

Hypothesis 2: Each parent contributes one factor-token to the offspring (so the offspring ends up with two factor-tokens.)

It is evident that hypothesis 2 postulates fewer causal processes than hypothesis 1: hypothesis 1 postulates the causal processes postulated by hypothesis 2 and another not postulated by hypothesis 2. Moreover, it would appear that hypothesis 2 can explain *at least* as much data as hypothesis 1. The only data that might be capable of being explained by hypothesis 1 but not by hypothesis 2 would appear to be the fact that the offspring of two different pure breeds all possessed the same characteristic. But this can be explained by saying that the characteristic "A" is dominant, a hypothesis we in any case need to explain why all members of Gen1(G) are yellow. So hypothesis 2 can explain at least us much as hypothesis 1, but it does so by postulating fewer causal laws than hypothesis 1. Therefore, incorporating hypothesis 2 into our overall explanation of the inheritance of characteristics results in a theory that is more independent of the data than does hypothesis 1. Hence the acceptance of hypothesis 2 is explicable on the view advocated here.

So, we have established that each parent contributes one factor-token to an offspring. But since every organism is also the offspring of its parents, it follows that every organism possesses two factor-tokens, one inherited from each parent.

Now, it might be objected that all the above argument shows is that each organism must possess *at least* two factor-tokens: the possibility has not been ruled out that some, or even all, organisms might possess more than two. However, it is easy to see that this hypothesis would contain explanatorily otiose components. Mendel found that aspects of the plants he studied were always one of only *two* ways: the pods were either green or yellow, the peas had either smooth skins or angular skins, the plants were either tall or short, and so on. He never found them to be some third way; for example, he never found brown pods, or skins covered in spikes and so on. So, given the data that were available to Mendel, the postulation of a third factor-token possessed by some or all plants played no explanatory role. Thus, on the view advocated here, the hypothesis that each plant had *exactly* two factor tokens is the hypothesis to be preferred.

So, in summary, it has so far been argued that the acceptance of the following two claims is explicable on the view advocated here:

1. Each organism contains within it exactly two factor-tokens.
2. In reproduction, an organism receives exactly one factor-token from each of its parents (and, of course, passes on exactly one factor-token to its offspring).

Now, let us consider the first generation of descendants from the crossing of the two pure breeds. It has been argued that every member of this generation have the factors "Aa". So the crossing of members of this generation with other members of this generation is in all cases the crossing of one plant with "Aa" with another plant with "Aa". Plainly, each offspring from such crossings will get either an "A" or an "a" factor from one parent and an "A" or an "a" from the other. There are, therefore, four possible ways in which an organism can obtain factors from their parents: an "A" from their mother and an "A" from their father, which can be denoted as "AA"; an "A" from their mother and an "a" from their father, which can be denoted by "Aa"; an "a" from their mother and an "A" from their mother, which can be denoted by "aA"; and an "a" from their mother and an "a" from their father, which can be denoted by "aa". There are, that is, four *possibilities* for the constitution of this generation, given by "AA", "Aa", "aA" and "aa".[6] The question arises: "What proportions of the next generation ought we to expect to have these various constitutions?" Here we may distinguish two ways in which the factors combine with each other to yield the four possibilities: either the process is essentially random, or it is not. If it is *not* random, then there must be some factor or mechanism M responsible for the particular way they combine. But Mendel had no evidence for the existence of such a mechanism M, and so postulating such an additional mechanism brings about an increase in the dependence of theory from data. Consequently, we obtain a more independent theory if we assert that they combine at random. But if they do combine at random, then we would expect, in the long run, that each of the four combinations will be more or less equinumerous.

It is plain that this last conclusion leads to the testable prediction that each possibility ought to make up about one quarter of any large body of descendants. But now, let us recall that of the two characteristics "A" and "a", the former is dominant, the latter recessive. Hence any plant containing "AA", "Aa" or "aA" will only exhibit, in their observable appearance, the dominant characteristic of yellow pods. Only those plants containing "aa" will have green pods. But in any large population, this will probably be about one quarter of the total. We are thus led to the following testable prediction: about one quarter of the plants obtained by crossing from within the first generation of plants descended from the two initial pure breeds will have green pods; the remaining three quarters will have yellow pods. And this is, of course, just what Mendel found.

There are a number of points that can be made about this result. First, note that we have derived the *prediction* that about one quarter of the plants will have green pods without using the observed fact that this is indeed the case. Therefore, the correctness of the prediction confirms the correctness of the reasoning that led to that prediction. Moreover, we can at this point use the bootstrapping AIM inference. One set of considerations has led us to the conclusion that the probability of any one combination of factors is precisely one quarter. But empirical observations have independently supported the suggestion that the probability of any one combination is about one quarter. Using the bootstrapping AIM inference, we can draw the conclusion that the probability of any one combination is, therefore, *exactly* one quarter.

And so the forms of inference advocated here can explain the acceptance of the basic principles of Mendelian genetics. These principles are as follows:

1. That the mechanism by which observable characteristics are passed from parents to offspring is by the inheritance of unobservable factors which are responsible for those observable characteristics.
2. The factors for the characteristics passed from parents to offspring always come in pairs, only one of which is ever manifested in the observable appearance of any particular organism.
3. For each such pair of factors, each organism receives one token of a factor from one parent and one from the other, so that each organism possesses exactly two tokens of the factors belonging to each pair.
4. In any pair, one of the factors is dominant, the other recessive, in the sense that if factor A is dominant and a recessive, the organism will display the observable characteristics associated with A.
5. The four possible combinations of factors AA, Aa, aA, and aa that can be found in any pair are equiprobable.

It has been argued that the acceptance of all five of these principles can be explained using the notion of the independence of theory from data, the preferability for low whole numbers and the bootstrapping AIM inference. And so the account given here also gives a probabilistic explanation of the empirical successes of these principles. In particular, it provides us with an explanation of the ability of these principles to successfully predict the frequencies of various characteristics in any number of subsequent generations obtained from some initial stock. This lends further support to our overall hypothesis that the apparatus developed here can explain the surprising successes of science we noted in Chapter 1.

CHAPTER 9
Conclusion

We have now completed our explanations of the three phenomena given at the start of this book. In Chapter 2 we laid down four criteria of adequacy that must be met by any such explanation. Let us now determine the extent to which the explanations offered here meet these four criteria.

THE FIRST CRITERION

A satisfactory explanation must explain how we have managed to hit upon successful theories

In Chapter 2 we noted two corollaries of this first criterion. The first corollary was the *accessibility requirement* that any property M of theories used to explain the forms of success must be more accessible that the forms of success it explains. The second corollary was the *explicability requirement* that any satisfactory explanation must explain why we have preferred theories with M rather than any one of the other highly accessible properties of theories.

Most of the work in explaining the phenomena has been done by the notions of the independence of theory from data and intra-DEC independence. In some cases we also explained the success of a theory by appealing to the fact that it used a low whole number, or that it was supported by the AIM inference. Plainly, all of these features of theories meet the accessibility requirement. A theory is independent of the data if it has a high ratio of components of data explained to dependent explanatory components of theory. As we have already observed, this is an easily accessible feature of a theory. The components of data that have been explained by a theory are determined by observation. A dependent explanatory component of theory is a restricted or unrestricted universal generalization couched in terms of straight predicates. But that something is such a generalization and whether the predicates used in it are straight are all easily determined. Therefore, the degree of independence of a theory is a highly accessible feature of the theory, and certainly more accessible than the phenomena we seek to explain. The same is clearly true of the forms of intra-DEC independence we discussed. Even more obviously accessible is the fact that a theory uses a low whole number or is supported by the AIM inference. So, the various properties of theories that we have used in explaining the forms of success clearly meet the accessibility requirement.

The explanations offered also seem to be able to meet the explicability requirement of explaining why we have preferred theories with *those particular* properties, rather than any of the many other accessible properties of theories. In Chapters 3–5 arguments were presented for the claim that if a theory has inductive support, a high degree of independence or intra-DEC independence, is a conservation law, uses a low whole number, or postulates a symmetry, it has an elevated probability of enjoying predictive success.

One of the main aims of Chapters 6–8 was to examine the *arguments* that Newton, Einstein and Mendel used in order to arrive at their theories. In the cases of Newton and Einstein, the arguments were fully and explicitly developed. In the case of Mendel, the argument was implicit and suggested rather than explicitly stated. We saw that the arguments by which these scientists arrived at their theories were designed to find that theory capable of *explaining* the data that also maximized the independence of theory from data.

For our purposes, it is important to note that Newton, Einstein and Mendel were *led* to their theories by the arguments they had constructed. The arguments led them to theories that either had a high degree of independence from the data or possessed some other property of which it was reasonable to say "this could not merely be due to chance". Since they were led to their theories by arguments, the theories they postulated were not mere "stab-in-the-dark" guesses.

The above considerations provide us with an explanation, meeting the explicability requirement, of why scientists prefer theories with *a high degree of independence* as a property of systems of DECs. Since the argument for the preferability of DECs with a high degree of *intra-DEC* independence is essentially the same, it also explains why they prefer theories with that property. The argument given for the preferability of theories that contain low whole numbers appealed to the fact that it was a consequence of all, or most, of the most likely explanations of why the measured value of some quantity should lie so close to a low whole number that the true value of the quantity should be the low whole number precisely. Now, if a scientist knows that it is a consequence of the best explanation E of why the measured value of a quantity should lie close to a low whole number that the true value of the quantity should be that low whole number precisely, and if the scientist knows that E is likely to have true empirical consequences, and the scientist wants true empirical consequences, then we have an explanation of why the scientist chose the value of the quantity to be that low whole number precisely. But what may still require an explanation is how the scientist came to know that E is likely to have correct empirical consequences. However, of course, the view advocated here shows that there are a number of ways in which the scientist could come to know this: by showing it was highly independent of the data, for example, or that it had a high degree of intra-DEC independence. So this reason for preferring theories with low whole numbers is certainly capable of meeting the explicability requirement.

The final type of inference discussed was the AIM inference. We make the AIM inference when we infer from the fact that two independent methods agree, and that their agreement was *a priori* highly unlikely, the conclusion that both methods have an elevated likelihood of being reliable in the future. It was argued that the inference *from* the *a priori* unlikely agreement of the two methods *to* an increased likelihood of their reliability is *a priori* plausible. And so, provided that the application of two independent methods *has* resulted in *a priori* unlikely agreement, we have an explanation, plausibly meeting the Explicability Requirement, of why scientists think those methods are more likely to prove reliable in the future. But what has not yet been explained is the *a priori* unlikely agreement of the two methods. How did we manage to hit upon methods that gave the same results? One *possible* answer is that those methods were warranted by theories that themselves had a high degree of independence from the data. Unfortunately, however, it seems at least rather questionable whether this reply will always be available to us. In some of the examples considered, particularly in Chapter 5, one of the methods used in arriving at, for example, a belief in *conservation* laws is *derivation from metaphysical principles*. If metaphysical principles are truly *a priori*, then no explanation has been given here of why it is they should succeed in leading to empirical predictions that are in close agreement with actual measurements. But one *possible* response to this might be to hold that metaphysical principles are not absolutely *a priori*, but enjoy a very close-to-*a priori* status similar to that of the conservation laws. *Provided* this is allowed, then the account given here seems to meet the Explicability Requirement.

THE SECOND CRITERION

A satisfactory explanation must also explain why it is that successful theories are, in fact, successful in the ways exemplified in the three phenomena

In Chapters 3–5 arguments were presented for the claims that theories with inductive support, high degrees of independence and intra-DEC independence, which use, under certain conditions, low whole numbers, or which are supported by the AIM inference, have an increased likelihood of enjoying subsequent predictive success. This provides us with a probabilistic explanation of why theories possessing these properties have, in fact, enjoyed rates of empirical success above that which would be expected purely by chance.

THE THIRD CRITERION

Any satisfactory explanation of the phenomena must not leave it merely as a fortunate fluke that the type of theory that scientists have preferred also happens to be the very same type of theory as that which (tends to) enjoy the forms of success exemplified in the phenomena

Chapters 6–8 were the historical chapters. In Chapters 6 and 7 it was argued that the *arguments* explicitly developed by Newton and Einstein make inferences to theories that maximize independence, or inferences defended in Chapter 5. In Chapter 8 it was argued that the theory choices made by Mendel can be reconstructed as attempts to increase independence, or as the result of the inferences defended in Chapter 5.

So those chapters together constitute a case for saying that Newton, Einstein and Mendel did seek theories that tended to maximize independence (and related notions) from the data. And since they did seek theories of this sort, they plausibly had some sort of implicit knowledge of the arguments showing that *if* a theory has independence, it is more likely to enjoy empirical success. A survey conducted by the author suggests that this may be true of scientists more generally.[1] Hence, on the view advocated here, it is not merely a fortunate fluke that the property of theories that explains their success is the very same property that these scientists look for in theories. The scientists we have examined, we may assume, want successful theories, they have (possibly implicit) knowledge that theories with independence are more likely to be successful, and so they choose theories with independence. But *those same arguments*, of which these scientists have implicit knowledge, also tell us that independent theories *are, in fact*, more likely to be successful. So we can explain why the property that is responsible for success is one and the same as that preferred by scientists.

THE FOURTH CRITERION

Any satisfactory explanation must be able to account for the actual historical examples of the phenomena

Of course, providing an explanation of every single instance in the history of science of the phenomena with which we are here concerned would be beyond the scope of any one book. The most that can be hoped for is a reasonably broad survey of some of the most important examples of the phenomena.

The first of the phenomena was the ability of science to successfully predict what seemed to be *novel* observations. We noted that the notion of a novel prediction has proved very difficult to define. Nonetheless, we will have explained everything that actually requires explanation if we succeed in explaining *actual cases of what we regard as* novel predictive success. So, for our purposes, we do not need a *definition* of novel success. In Chapters 6–8, explanations were developed of a range of cases which we would certainly be inclined to regard as cases of novel success. Our strategy was to show that the theories used in the derivation of the novel phenomena were highly independent of the data. Since it is a consequence of arguments developed in Chapters 4 and 5 that theories that have a high degree of independence are more likely to enjoy predictive success, we thereby have a probabilistic explanation of predictive success of those theories.

In Chapter 6 it was argued that Newton's three laws of motion and law of gravitation were highly independent of the data. Moreover, these laws have an essential role in the derivation of a wide range of novel predictions, including the prediction of the outer planets Neptune and Pluto, their positions, orbit paths, momenta and departures from perfect sphericity. In Chapter 7, the special theory of relativity was considered. It was argued that it, too, possesses a high degree of independence from the data. But the special theory of relativity has an essential role in the derivation of some novel, and very surprising, phenomena, including the conversion of mass in to energy that takes place in the atom bomb and in nuclear reactors, the slowing of extremely accurate moved clocks, and the extended life-span of particles accelerated close to the speed of light. In Chapter 8 it was argued that Mendel's theory of genetics was highly independent of the data. This gives us an explanation of the ability of the theory to accurately predict frequencies of characteristics in many generations of offspring. So the account offered here explains some apparently novel successes in physics and biology.[2] While this is a fair range of cases, I acknowledge that it does not deal with either general relativity or quantum theory. The latter, of course, is perhaps the most predictively successful theory in the history of science. The reason for these omissions is simply that I do not know much about either of these theories. I do, however, intend to work on them in the future.

The second phenomenon was the ability of science to lead to knowledge of areas that were not accessible at the time theories about those regions were first postulated. There are only a few *uncontroversial* cases of this phenomenon. One example that was given in Chapter 1 was the ability of Newtonian physics to predict a planet – that is, a very large spherical material object – beyond the orbit of Uranus. It was argued in Chapter 5 that the view advocated here can explain this example.

The third of the phenomena with which we have been concerned is the ability of some *a priori* or close-to-*a priori* theories to lead to empirical predictions that were subsequently confirmed. Chapter 6 discussed a number of examples of this phenomenon; in particular, in that chapter it was argued that this account can explain both the close-to-*a priori* character of some aspects of Newton's laws of motion and their empirical success. So we may conclude that we have given explanations of some examples of all three kinds of the surprising phenomena we examined at the start of the book.

THE RELATION OF THE EXPLANATION OF THE PREDICTIVE SUCCESS OF SCIENCE ADVOCATED HERE TO THE HYPOTHESIS OF SCIENTIFIC REALISM

In this book a way of explaining the ability of science to successfully predict novel phenomena has been developed. The explanation does not appeal to truth, or truth-likeness of any of the theoretical parts of the successful theories. Neither does it appeal to the reality of any theoretical entities. That is, it offers an explanation of the ability of science to predict novel phenomena that does not employ the devices typically used by the scientific realist. And so the questions naturally arise: does the position advocated in this book undermine the claim that scientific realism is the only philosophy of science that does not make the success of science "a miracle"? Does it render unsound the argument for scientific realism from the novel success of science?

In this section it will be argued that, although the conclusions defended in this book show that the arguments for scientific realism from novel success may need to be recast, they do not render unsound all such arguments. More specifically, it will be argued there is a distinction between, on the one hand, explaining the ability of a theory, or scientist, to *successfully predict* a novel phenomenon and, on the other hand, explaining the novel phenomenon *itself*. Scientific realism, it will be argued, is not required to explain the former, but may be required to explain at least some forms of the latter. We can get a clearer grasp of these issues if we consider a specific example. Let:

E_1 = the ability of Fresnel's theory to successfully predict the existence of a white spot in the middle of a perfectly circular shadow.

And let:

E_2 = the existence of the white spot in the middle of a perfectly round shadow.

E_1 refers to an ability to predict a phenomenon; E_2 refers to the phenomenon itself. It will here be argued that an explanation of E_1 need *not* also be an explanation of E_2.

Let us begin by noting that an explanation must do more than show that the explanandum can, with some degree of probability, be expected. After all, the appearance of "Koplik spots" in the mouth of a patient enables us to assert, with high probability, that the patient will within a few days develop the symptoms of measles. But we would not say that the Koplik spots *explain* the measles. Koplik spots may enable us to predict the measles, but we would not generally say they enabled us to explain the measles. To explain the measles we would, perhaps, need to make reference to the presence of the measles virus in the patient.

Nevertheless, there is, perhaps, another, related phenomenon that Koplik spots might enable us to explain. Let us imagine that we are members of a community that knows nothing of Koplik spots or of the measles virus. No one in our community knows about the measles virus, and neither (we are assuming) do we. But every now and again a member of the community develops measles. We note that a local "medicine-man" is able to successfully predict the appearance of measles in a patient. No one else in the community is able to do this. We want to know how the medicine-man is doing it. We find out the following: that the medicine-man has discovered that in the past measles has been preceded by the appearance in the patient's mouth of white spots (Koplik spots). The medicine-man has observed many instances of this, and this has given him confidence this regularity will continue to hold in the future. He then uses the appearance of Koplik spots in a patient to predict that the patient will develop measles a few days later. But, the medicine-man does not know about the measles *virus*. All he knows is the statistical link between Koplik spots and the measles *symptoms*.

Plausibly, we would not now be in possession of an explanation of measles: all we know is the statistical link between the Koplik spots and the other symptoms of measles. We are ignorant of the measles *virus*. But would we say we had succeeded in explaining *how the doctor was successfully predicting the appearance of the measles symptoms a few days later*? I believe we would. And this is so even though we had not obtained an explanation of the measles symptoms themselves. *Explaining the ability to successfully predict an event E is not the quite the same thing as explaining E itself.*

It might be objected the position just defended is not logically coherent. The assertion "the doctor successfully predicted that the patient would get measles" clearly entails "the patient got measles". Since the former assertion logically entails the latter, how can something be an explanation of the former without also being an explanation of the latter? And since the account given of how the doctor came to successfully predict the measles plainly does not explain the measles themselves, it might be argued that the account given therefore does not *explain* how the doctor came to successfully predict the measles either.

The foregoing objection is based on the following principle concerning explanation and logical entailment:

If E explains O, and O entails O*, then E also explains O*. (ELE)

Although this principle might seem plausible, it will be argued it is false. In particular, it will be argued it is false when what is explained is a "cognitive achievement". It is also not in general true when explaining in to "intentional contexts".

We can illustrate the falsity of ELE by considering a simple example. Let:

O = "Jim knows the chocolate biscuits have disappeared."
O* = "The chocolate biscuits have disappeared".

An explanation of *how Jim knows* the chocolate biscuits have disappeared might be along the following lines: "Jim clearly remembers placing the chocolate biscuits on the shelf before he left home. When he returned home they were gone. He looked everywhere in the house but could not find them. No one confessed to taking them." This is clearly an explanation of *how Jim knows* the chocolate biscuits have disappeared, but it in no way explains the actual disappearance of the biscuits themselves. And this is despite the fact that "Jim knows the chocolate biscuits have disappeared" entails "the chocolate biscuits have disappeared". E can explain O and O entail O* without E explaining O*. The principle ELE therefore seems to be false.

Another example confirms the falsity of ELE. Suppose Smith has successfully predicted the outcome of a horse race, and we wish to explain how he has done it. One explanation might be as follows: Smith has overheard two "colourful racing identities" discussing a particular race. They both confidently agree that a notoriously slow horse, with long odds, will win a particular race. Suppose the two identities have a remarkable good record of successfully picking winners. This leads Smith to reason that since these two identities have been good at picking winners in the past, the horse they are discussing will win. And suppose it does win. Then we would, I think, have an explanation of how Smith successfully predicted the outcome of the race. But I don't think we would say we had an explanation of why the horse actually won. And this is despite the fact that "Smith successfully predicted the horse will win" entails "the horse will win". So I think we may conclude that the principle ELE, despite initial plausibility, is false.

The examples given also show, I think, that an explanation of some cognitive achievement, such as knowing that P or successfully predicting that P, need not also be an explanation of P itself. Consequently, there seems to be no bar to saying that the view developed here gives us an explanation of how scientists have managed to successfully predict novel phenomena, even if it does not constitute an explanation of the phenomena themselves.

THE "NO MIRACLES" ARGUMENT FOR SCIENTIFIC REALISM

Hilary Putnam argued that scientific realism is the only philosophy of science that does not make the success of science "a miracle" (Putnam 1979: 73). On the face of it, it might appear that the view advocated here is incompatible with Putnam's "no miracles" argument; after all, here an *alternative* explanation of predictive success, not making use of the devices of realism, is offered. On this view, a theory's predictive success is to be explained by its high degree of independence from the data, and not by of its truth or any kindred property. It might *appear*, therefore, that the view advocated here obviates realist explanations of success. But this is not so. It follows from the previous section that the view offered here *need not* undermine the arguments for scientific realism from the success of science.

We have already noted that an explanation of, for example, how the medicine-man predicts the Koplik spots need not be an explanation of the Koplik spots themselves. We can explain how the medicine-man successfully predicts the Koplik spots by saying he has observed that, in the past, a correlation has held between Koplik spots and measles. But to explain the measles themselves, another hypothesis is needed. That other hypothesis might, for example, concern the presence of the measles virus in the patient. Similarly, here it has been argued that we can explain cases of novel predictive success in science by saying that scientists have observed certain patterns in the data, that those patterns are highly independent of the data and that they therefore have an increased chance of being observed in the future. This, it has been argued, explains *the ability of the scientists to successfully predict* the phenomena, but it need not constitute an explanation of the phenomena themselves. The possibility remains that an explanation of another sort is required to explain the phenomena *themselves*. And it is here that there may still be possible to construct an argument for scientific realism from the success of science.

THE SCIENTIFIC REALISM DEBATE RECAST

In this section it will be argued that the cogency or otherwise of the arguments for realism, and the arguments for and against the various forms of realism, remain more or less untouched by the perspective developed here. We have noted that an explanation of a scientist's ability to successfully predict a phenomenon need not also be an explanation of the phenomenon itself. And so the question arises, "What would constitute an explanation of the phenomenon itself?" A natural answer to this is "The actual laws used in explaining the phenomenon." What would, for example, provide us with an explanation of the white spot in the middle of the round shadow? A natural answer is "The laws of optics proposed by Fresnel would constitute such an explanation." More generally, it is the *laws of science* that provide us with explanations of the novel phenomena themselves.

On the face of it, this simple observation would seem to have no bearing on the issue of scientific realism. But an obvious point can also be made here. If a theory is interpreted wholly instrumentally, it clearly cannot provide us with what we would ordinarily regard as an *explanation* of anything. An explanation must, it seems, at least show how an observation-statement can be logically derived from a theory, and it perhaps also ought to give us an understanding of how something came to be, or answer why things are one way rather than another. But to do any of these things, a theory must at least be *meaningful*. It cannot do any of these things if it is interpreted wholly instrumentally. But, of course, this point is not sufficient to establish scientific realism. Even if a theory needs to be *interpreted* non-instrumentally if it is to provide us with explanations at all, it does not follow that we are justified in *accepting* the theory, or *believing* it to actually be true. The theory might be perfectly meaningful, but no more credible than fiction. And so the following questions arise. Given that a theory T must at least be meaningful to explain some phenomenon P, does the fact that it explains P make it worthy of rational acceptance? Are we entitled to say the entities postulated by it exist? Is the case for its acceptability especially strong if P is a novel phenomenon? Are we, perhaps, only entitled say T is right about the structure, or that the world is "as if" T is true? That is, the questions that arise are the questions of the scientific realism debate. So, even if it is possible – as has been argued in this book – to explain how science is predictively successful without using the devices of the scientific realist, still the questions of scientific realism arise. But the issue is perhaps not best expressed as "Is scientific realism the best explanation of the predictive success of science?" Better formulations of the question might be "Does the predictive success of a theory justify a scientific realist attitude to it?" or "Does the success of the theory perhaps only justify a scientific realist view of it under certain conditions, or to only some parts of the theory?". But the issue of scientific realism remains.

CONCLUDING REMARKS

Throughout this book we have been concerned with explaining certain types of success in science. These types of success are the ability to successfully predict novel phenomena, the ability of some theories to give us knowledge of parts of reality not accessible at the time those theories were initially postulated and the surprising success of some close-to-*a priori* scientific theories. All these cases of the success of science involve science giving out surprisingly more true empirical information than we have put in to it. In all these cases, it seemed, science somehow produced the rabbit of surprisingly novel empirical knowledge from the hat of previously obtained data. Our aim in this book has been to explain how it has done this. The key notion used is the independence of theory from data. Scientists look for patterns in obtained data, and in any actually obtained body of data there will be indefinitely many such patterns. But, it has been argued, the less ad hoc dependence a pattern has on the data, the better the chances the data will continue to conform to that pattern. Moreover, if a pattern is highly independent of the data, that the past data and the future data do indeed exemplify *the same* pattern may be far from obvious. It can appear as though data of an entirely novel kind are being predicted. And so we have at least a *possible* explanation of the ability of scientists to successfully predict phenomena apparently quite different from those initially used to formulate their theories.

In the previous three chapters it has been argued that this possible explanation of the surprising success of science applies to Newton's description of the solar system, Einstein's special theory of relativity and Mendel's theory of genetics. The notion of independence of theory from data can be used to explain actual cases of surprising success from the history of science. But at the start of this book we noted that the ability of science to be successful in these surprising ways seemed to at the core of why we regard science as our best example of rationality. The view developed in this book, therefore, also offers an answer to the question "What's so great about science?".

Notes

1. SOME SURPRISING PHENOMENA

1. This expression was first used by Robert Junk (1956). It was subsequently used by Hilary Putnam in his "Degree of Confirmation and Inductive Logic" (1979).
2. Actually, the predictions of the time gain of the clock involved both the Special Theory and the General Theory of Relativity.
3. For a more comprehensive list of cases of novel predictive success in science, see T. Lyons's "Scientific Realism and the Pessimistic Meta-Modus Tollens" (2002).
4. Pauli first postulated the neutrino in a letter written in 1925, addressed to a number of physicists, which is reproduced in *Collected Scientific Papers by Wolfgang Pauli* (Kronig & Weisskopf 1964). The discovery of the neutrino was first announced in a letter to the Editor of *Physical Review* (Reines & Cowan 1953).
5. A clear and comprehensive account of Eddington's observation of stars near the Sun during an eclipse, and the relation of those observations to the general theory of relativity, is given in *Was Einstein Right?* by Clifford M. Will (1946, esp. pp. 65–88). It is now known that Eddington actually selectively used his data so that his results appeared to lend more confirmation to Einstein's data than they actually did. However, of course, later observations have confirmed Einstein.
6. Confirmations of this effect have even been found by observing a "blue-shift" in the light emanating from objects at the top of a tower, as compared to light emanating from objects at the bottom of the tower. The strength of the Earth's gravitational field at the top of a high tower is very slightly weaker than at the bottom. The general theory of relativity leads to the prediction that the light emanating from objects at the top of the tower will therefore be very slightly shifted to the blue end of the spectrum. This has been observed to be the case in the Pound–Rebka experiment.
7. The discovery of the W and Z was announced in "Experimental Observation of Isolated Large Transfer Energy Electrons with Associated Missing Energy at s = 540GeV" (Arnison *et al.* 1983). A popular account of this discovery is given in P. Watkins, *The Story of the W and Z* (1986).
8. It might be thought that a natural way to mark the distinction between inductive and ontological grounds for doubting a theory is as follows. Let a theory be of the form "All A are B". Then we have inductive grounds for doubting the theory if we believe there might be an A that is not a B, and we have ontological grounds for doubting it if we believe that there may really be no A. However, this will not do. The fact that there are no inertial systems would not seem to be a reason of any kind for doubting the truth of either Newton's first law of motion or the special theory of relativity. Therefore, the non-existence of such systems cannot be an *ontological* ground for doubting either of those theories. Rather, we have ontological grounds for doubting a theory T if it says that some observed phenomena are due to some entities *E*, and we have reasons to believe that there are no such entities.

9. Although the *conceptual* distinction between inductive and ontological grounds for doubting a theory is quite clear, there are some cases in the history of science which could be seen as giving us either inductive or ontological grounds for doubting a theory. One possible example concerns the relationship between the special theory of relativity and Newtonian dynamics. Newton thought that the mass of any object was always independent of its velocity. The special theory of relativity could be seen as either giving us inductive reasons for doubt concerning this claim, by asserting that mass is dependent on velocity, or ontological grounds, by asserting that there really is no such thing as Newtonian mass.

10. For example, Kant thought that Newton's laws were deducible from *a priori* principles. James Clerk Maxwell also thought Newton's laws were *a priori*. It is worth noting that Newton's three laws can be seen as conservation laws, specifically relating to the conservation of momentum.

11. To say that a theory T is *a priori preferable* is not the same as to say it is *a priori plausible*. To say that a theory T is *a priori* preferable is to say that it is, *a priori*, to be preferred to some other theory or theories. But to say it is *a priori* plausible is not in itself to say anything about its relation to other theories. Nonetheless, the two notions are connected: to say that a theory T is *a priori* preferable to another theory T* is to imply that T has *more a priori* plausibility than T*. Therefore, to say that T is *a priori* preferable to T* implies that T has at least some *a priori* plausibility, although this amount of *a priori* plausibility need not be by itself enough to justify T's acceptance.

12. See, for example, "Experimental Gravitation from Newton to Einstein" by C. M. Will (1987). Will notes both that there has been theoretical speculation that gravity does not obey an inverse square law, and also some, although inconclusive, experimental evidence that at short distances it does not.

13. Neither the strong nor weak subatomic forces obey an inverse square law. See, for example, P. Watkins, *The Story of the W and Z* (1986).

2. SOME UNSATISFACTORY EXPLANATIONS OF THE PHENOMENA

1. For a dissenting view, see Robin Collins, "Against the Epistemic Value of Prediction Over Accommodation" (1994).

2. These results come from S. D. Drell, "Experimental Status of Quantum Electrodynamics" (1979). Drell also mentions impressive confirmations for QED with respect to the predicted values of the "first excited states of positronium", the "helium fine structure", "muonic X-rays" and the "Lamb shift in μ-helium".

3. I do not know the range of results that could have been obtained by the measuring apparatus used to calculate the magnetic moment of the electron, but here I have assumed that, from an *a priori* point of view, the results could have been anywhere between zero and $2000000000 \times 10^{-12}$. Since it is *a priori* possible that the obtained result could have been greater than the latter figure, the estimate of the *a priori* probability that this method gives is bound to be an *over*estimate.

4. One well-known representative of this view is Richard Boyd. His views are developed in his "On the Current Status of the Issue of Scientific Realism" (1983).

5. One difficulty, of course, is *defining* the notion of simplicity. This is a notoriously difficult problem.

6. Structural realism is defended by John Worrall in "Structural Realism: The Best of Both Worlds" (1989).

7. Stathis Psillos (1999), however, has argued against the claim that we can draw a sharp distinction between the (relatively more) knowable and (relatively less) knowable nature.

8. This is not to deny the point, made by some structural realists, that knowledge of structure might be *more accessible* than knowledge of nature. It is simply to say there is still a problem about how we come to know structure, where that structure is sufficient to make novel predictions.

9. A prominent advocate of a broadly meta-inductive approach to epistemology is Philip Kitcher. He gives a detailed development and defence of a naturalised approach to epistemology in his "The Naturalists Return" (Kitcher 1992).

10. I owe this objection to Robert Nola.

11. A similar point is made by John Worall in his paper "Two Cheers for Naturalised Philosophy of Science – Or: Why Naturalised Philosophy of Science is Not the Cat's Whiskers" (1999).

12. One philosopher who has offered a broadly "evolutionary", or trial and error, account of the success of science is Larry Laudan. See, for example, his "Explaining the Success of Science: Beyond Epistemic Realism and Relativism" (1984).

13. A very useful survey of these matters can be found in *Discovery, Creativity and Problem-Solving* by David Lamb (1991).

14. A "neural net" can be trained to perform a task, such as recognizing objects of a certain sort simply by being told when it has recognized a given object correctly and when it has not, and modifying its own internal workings in the light of this information. It can in this way acquire the ability to reliably recognize objects of a particular sort, but it is not following an algorithm or set of instructions, except in a rather trivial sense.

15. It is also worth noting here that Poisson, who was the first person to derive the prediction of the white spot from Fresnel's wave theory of light, did not believe the correctness of that theory himself. He was actually trying to *disprove* it. He thought that when he had shown that it was a consequence of Fresnel's theory that there would be a white spot in a circular shadow, he had succeeded in finding a *reductio ad absurdum* of the theory. But when the experiment was performed it was found, contrary to Poisson's expectations, that the white spot did exist.

16. See, for example, Lakatos's plausible remarks on the failure to find a predicted planet in his "Falsification and the Methodology of Scientific Research Programmes" (1970).

3. A DEFEASIBLE *A PRIORI* JUSTIFICATION OF INDUCTION

1. This is so if we interpret "induction" *broadly*, that is if we interpret "inductive inference" to be equivalent to "ampliative inference". If we interpret "induction" more narrowly, as, for example, the straight rule (SR), then SR must be justified by some ampliative inference other than the straight rule. But then the question arises how I is to be justified, and so we are back to square one.

2. This is not the place for an extended defence of the notion of the *a priori*. However, such an extended defence can be found in BonJour's *In Defence of Pure Reason* (1998). BonJour has two main arguments for the existence of the *a priori*, including the synthetic *a priori*. The first argument is from the existence of convincing examples of *a priori* belief. The second (and in my view more significant) argument is that, unless we allow that there can be non-empirical evidence, we can never have any reason to believe anything beyond "observation sentences", however, precisely, they are characterized. This second argument, if sound, establishes the existence of synthetic *a priori* statements or inferences that are ampliative but *a priori* reasonable, or both.

3. In this respect, the view of *a prioricity* advocated here *may* be a little different from that advocated by BonJour. For BonJour, *a priori* insight is always the insight that things *must* be a certain way, and hence that the relevant *a priori* claim is *necessarily* true. But, for BonJour, an *a priori* claim may nonetheless be defeasible. For BonJour this implies that our insights into how the world must be might themselves be mistaken.

It is worth noting that the view advocated here perhaps need not be incompatible with BonJour's. Consider the conditional:

If the probability of E is 1, then E will occur. (1)

This is not a necessary truth. But perhaps the following claim, or something like it, is necessary:

It is rational to believe (*ceteris paribus*) that if $Pr(E) = 1$, then E will occur. (2)

Plausibly, (2) is something that is just part of what we mean by "rationality". Moreover, it is I think plausible to say that it is (2) rather than (1) that is more immediately the product of *a priori* insight. But, if so, then (1) can be something that is rational to believe as a result of *a priori* insight, and hence – provided the other conditions for knowledge are met – (1) can count as *a priori* knowledge. And this can be so even if all knowledge that is *directly* the product of *a priori* insight is necessary.

4. It might be wondered what content could be given to the claim that the coins are all fair if, when an infinite number of them are tossed, they all come up heads. One natural way would be to appeal to a propensity interpretation of probability, and say that all the coins were fair in virtue of the fact that they were all disks with a perfectly symmetrical internal distribution of mass.

5. Like (7), (8) can be an *a priori* reasonable or justified belief, provided that we believe it with the appropriate degree of "epistemic modesty". Consider the claim:

$Cr(If Pr(E) = n$, then E will occur$) = n$.

A straightforward consequence of this is:

$Cr (If Pr(E) = n$, then E will *not* occur$) = 1 - n$.

This implies that if the probability of an event E is very low, it is reasonable to believe that E will *not* occur. Moreover, by the reasoning given above, the belief that "if the probability of E is some very low number n, then E will not occur" can be a belief that is synthetic but *a priori* justified if the strength with which we believe the conditional is not stronger than $1 - n$. This seems to show that (8), too, can be a synthetic but *a priori* reasonable belief, provided that it is believed with appropriate "epistemic modesty".

6. This is a somewhat simplified version of BonJour's argument. BonJour actually discusses the inference from "m/n observed A have been B" to "m/n A are B".

7. BonJour discusses "Schrodinger's cat"-type cases from quantum theory. See BonJour (1998: 212).

8. I think we regard the notion of a propensity for black things to be crows as bizarre because, while we accept that the genetic make-up of an organism could causally determine its colour, we also believe an organism already has a genetic make-up by the time it has a colour. So to postulate a propensity for black things to be crows is to place the effect before the cause.

9. The question of whether a purely statistical or probabilistic explanation really is satisfactory for our purposes is discussed in Chapter 7 (pages 154–8).

4. THE INDEPENDENCE OF THEORY FROM DATA

1. This remark is only "very rough" because we have not yet addressed the problem of individuating dependent explanatory components of theory and components of data.

2. The question arises, "Is it reasonable to suppose that mind- and language-independent relations of similarity exist?" This, of course, is a large philosophical question in its own right. It is perhaps sufficient here to note that there are at least two sorts of argument for their existence. One type of argument, developed in detail by Armstrong (1978), is that to say that relations of similarity *are* dependent on language leads to a vicious infinite regress. Another argument, or class of arguments, has been developed by Lewis, who has argued that the theory can do a great deal of philosophical work, and is therefore extremely useful. This can perhaps be seen as a kind of "abductive" argument for the existence of such relations. So there seems to be at least a defensible case for saying that the notion of mind- or language-independent similarity is defensible.

3. One way in which bivalence could fail even though there are mind-independent relations of similarity and difference is as follows. Suppose A and B are classes of objects, and they have no common elements. Suppose also that the members of A are objectively similar, as are the members of B. Let predicate P partially refer to A and to B. Finally, suppose object a is a member of A but not B. Then "a is P" will be neither true nor false. (It will be partially true and partially false.)

4. Saying that there are mind-independent relations of similarity and difference can be compatible with allowing "equivalent descriptions". Suppose we have two "cognitively equivalent" versions of Newtonian physics: one says masses are points, the other says they are fields centred at those points. This can be compatible with saying that the relation of similarity that holds between all masses is something that is mind- or language-independent.

5. It is weaker than Armstrong's realism about universals because it only requires us to say that some statements of the form "a is similar to b" are true independently of the mental, not that any universal *exists*, whether independently of the mental or otherwise.

6. Chaitin (2006) claims that a theory with a low Kolmogorov–Chaitin complexity gives us understanding.

5. SOME MORE SUCCESS-CONDUCIVE PROPERTIES OF THEORIES

1. In this argument we have arbitrarily restricted the range of possible values of the parameters A, B, C, etc. to whole numbers between +20 and −20. Of course, essentially the same point could be made using some other restriction on the possible values of the parameters, provided that the number of possible values they could take was finite. But, it may be objected, why place any restriction at all on the possible values? And if they are permitted to take any value, then the number of rules like One and Two will be infinite. And in such circumstances the argument for the preferability of Rule One will fail to go through. We can, I think, overcome this difficulty by introducing the notion of an *experimentally indistinguishable class of rules*. Let A be some parameter, and suppose our experimental techniques for determining the value of A are restricted in two ways. First, let us suppose, our apparatus cannot measure values of A above or below some number (say 100), since such values are simply "off the scale". This means that A = 110, for example, is *"experimentally indistinguishable"* from, say, A = 170. Second, we will suppose our apparatus can only measure values of A to a finite degree of accuracy. So A = 28 may be experimentally indistinguishable from A = 28.01. Then the number of experimentally indistinguishable sets of theories capable of explaining any given body of data will be finite. The argument given can then be used to show that it is *a priori*

less likely that there should be some experimentally indistinguishable set of theories with the same number of parameters as rule 1 that can explain the data than that there should be some experimentally indistinguishable set of theories with the same number of parameters as rules 2 or 3. Of course, modified in this way, the argument only gives us a reason to prefer a class of experimentally indistinguishable theories, rather than any one specific theory. But the arguments given in the next section, on the preferability of low whole numbers in science, can then be used to single out one particular member of a class of experimentally indistinguishable theories as being the most preferable.

2. Elsewhere, in *Science and the Theory of Rationality* (Wright 1991, esp. ch. 6) I have argued that the view developed here can apply to a comparison of Lavoisier's oxygen theory of combustion and the phlogiston theory of Cavendish. These are both theories that are difficult to represent as having "free parameters". But it was argued that it can be shown that the oxygen theory of combustion has a higher degree of independence from the data than phlogiston theory.

3. This thesis is further developed in Chapter 6, where it is noted that Newton's laws of motion can be seen as conservation laws.

4. In gravitational lensing, the gravitational field of a massive object acts as a lens, splitting the light from a distant source into two distinct beams.

5. If a point can lie anywhere on a number line, the *a priori* probability that it should be close to a large whole number is very much greater than the probability that it should be close to a low whole number, since there are very many more large whole numbers than low whole numbers.

6. NEWTON'S LAWS OF MOTION AND LAW OF GRAVITATION

1. Perhaps the best-known advocate of the thesis that Newton's laws are, in a sense, *a priori*, is Kant. However, it must be emphasized that no claim is made here that any of Newton's laws are indeed fully *a priori* in any sense. It is only claimed that they have a "close-to-*a priori*" status.

2. This statement of Newton's Laws comes from *Sir Isaac Newton's Mathematical Principles of Natural Philosophy and His System of the World* (1960), hereafter referred to as *Newton's Principia*.

3. As does, for example, Clark Glymour in his discussion of Newton's argument for universal gravitation. See Glymour's *Theory and Evidence* (1980).

7. SPECIAL RELATIVITY

1. It is not entirely clear whether Einstein really does subscribe to verificationism as a theory of the nature of truth: he repeatedly uses the word "true" in "scare quotes", apparently indicating that he finds the notion of truth to be somehow problematic or dubious. It is also worth noting that on p. 140 of *Relativity* Einstein gives an account of the origins of our notion of a "real external world" which, plausibly, brings him rather close to phenomenalism. He writes:

> Thus it comes about that A associates with B the experience "it is lightning". For person A the idea arises that other persons also participate in the experience "it is lightning". "It is lightning" is now no longer interpreted as an exclusively personal experience, but as an experience of other persons (or eventually only as a "potential experience"). In this way arises the interpretation that "it is lightning",

which originally entered into the consciousness as an "experience", is now also interpreted as an (objective) "event". It is just the sum total of all events that we mean when we speak of the "real external world".

Here Einstein starts off giving an account of the *origins* of our notion of an objective event, but he seems to end up giving something like a phenomenalist *analysis* of our notion of the "real external world".

2. Here will only concentrate on the reasons Einstein gives in *Relativity*. However, in his 1905 paper "On the Electrodynamics of Moving Bodies", the reason for the principle of relativity to which he gives most prominence is one coming from a particular phenomenon in electrodynamics. It had been observed that when a magnet and a conducting wire are moved relatively to each other, an electric current is produced in the wire. But Maxwell produced two *different* explanations of this phenomenon: one for when the magnet was taken to be at rest and the wire moving and another, quite different explanation for when it was the wire that was taken to be at rest and the magnet moving. But, Einstein notes, Maxwell has here given us two distinct explanations for what seems to be *in reality* indistinguishable observable phenomena. So, Einstein concludes, this case does not really constitute an example in which the phenomena of nature themselves suggest there is a difference between absolute rest and absolute motion. He goes on to state that this supports the principle of relativity. Presumably Einstein's argument here is that the only apparent *case* of the notions of absolute rest and absolute motion playing an explanatory role is the two explanations given by Maxwell of the current induced in a wire when it moves in relation to a magnet, but since this case is not really a case of the phenomena of nature indicating a difference between absolute rest and motion, it actually does not suggest that the laws of nature are dependent on what is at rest and what is in motion. Hence, we have *no* evidence of such dependency. So the hypothesis that there is such dependency is to be rejected and the principle of relativity accepted.

That Einstein should respond in this way to this phenomenon of electromagnetism is compatible with the view advocated here. Maxwell offered us two distinct explanations for what seems to be but a single phenomenon. His explanation also postulates something – a distinction between absolute rest and absolute motion – for which there is no evidence. Hence, Maxwell's account *increases* the dependence of our theory on the data. We should, therefore, reject the distinction between absolute rest and motion and seek a unified explanation of this phenomenon, thereby bringing about an increase in the independence of our theory. But this entails at least provisional acceptance of the principle of relativity.

3. It might be objected that this claim is a little dubious. Surely, it may be objected, if (T) is to make an analytically true claim about time intervals, it must be the case that the clocks are *accurate*, that is, if the clocks say one hour has passed, it must be the case that it is indeed one hour, and not 59 or 61 minutes, that has passed.

4. Einstein gives his derivation on pp. 115–120 of *Relativity*. Another very accessible derivation, but based on discussions of a rather different physical system (a "photon clock"), is given in Wesley Salmon's *Space, Time and Motion* (1975).

8. MENDELIAN GENETICS

1. Mendel himself does not use the expression "pure breed". He rather speaks of strains that "possess constant differentiating characteristics". See his "Experiments in Hybridization" (1913). The expression "pure species" does however occur on p. 373.

2. Mendel actually studied seven variable characteristics of the peas: (1) The difference in the form of the ripe seeds, (2) The difference in the colour of the seed albumen, (3) The difference

in the colour of the seed coat, (4) The difference in the form of the ripe pods, (5) The difference in the colour of the unripe pods, (6) The difference in the position of the flowers and (7) The difference in the length of the stem.

3. The following "rational reconstruction" of Mendel's argument is suggested in Mendel (1913), especially pp. 335–50 and 356–63.

4. This notation first appears at Mendel (1913: 349).

5. Mendel (1913: 346) writes: "If now the results of the whole of the experiments be brought together, there is found, as between the number of forms with the dominant and recessive characters, an average ratio of 2.98 to 1, or 3 to 1." Interestingly, Mendel did not *argue* for the legitimacy of his move from "2.98" to "3", but perhaps he felt he could not do so without begging the question in favour of the theory he wished to defend. It is also worth noting that it is this tiny, almost insignificant step that plays a crucial role in the derivation of one of the few theories outside physics to enjoy a greater degree of empirical accuracy than the data on which it is based.

6. Mendel himself adopts the useful notational device of representing the factor obtained from the pollen (male) cells and egg (female) cells as a fraction whereby the factor from the pollen is the numerator and the factor from the egg the denominator. He thus represents the four possibilities as A/A, A/a, a/A and a/a.

9. CONCLUSION

1. The author conducted a survey of a number of practising scientists at the University of Sydney. The scientists were given an abbreviated version of the argument for the preferability of independent theories given in Chapter 5. They were then asked if this was the rationale on which they made the choice between competing theories, all of which were able to accommodate the data. Of the eight scientists surveyed, seven agreed that it was. The one dissenter (an X-ray crystallographer) said that in his field the situation did arise where it was necessary to choose between two equally good theories: generally there was one theory that was the best. However, he agreed that if, hypothetically, he were faced with such a choice, the criterion given (independence from the data) is the one he would use.

The scientists interviewed were Dr Iver Cairns, Professor Richard Hunstead, Dr Tim Adams, Dr John O'Byrne and Associate Professor Rod Cross of the Department of Physics at the University of Sydney, Dr Julian Fernandez and Dr R. Fenton of the Department of Chemistry at the University of Sydney, and Dr Charles Collyer of the Department of Biochemistry at the University of Sydney.

2. I have elsewhere argued (Wright 1991) that the account can also explain the "chemical revolution", that is the transition from Priestley's version of the phlogiston theory of combustion to Lavoisier's oxygen theory of combustion.

Bibliography

Armstrong, D. 1978. *Universals and Scientific Realism*. Cambridge: Cambridge University Press.

Arnison, G. *et al.* 1983. "Experimental Observation of Isolated Large Transfer Energy Electrons with Associated Missing Energy at $\sqrt{s}$ = 540GeV". *Physics Letters B* **122**: 103–16.

Barnes, B. 1974. *Scientific Knowledge and Sociological Theory*. London: Routledge.

BonJour, L. 1998. *In Defence of Pure Reason*. Cambridge: Cambridge University Press.

BonJour, L. 2005. "In Defence of the *A Priori*". In *Contemporary Debates in Epistemology*, M. Steup & M. Sosa (eds), 98–104. Oxford: Blackwell.

Boyd, R. 1983. "On the Current Status of the Issue of Scientific Realism". *Erkenntnis* **19**: 45–90.

Brueckner, A. 2001. "BonJour's *A Priori* Justification of Induction". *Pacific Philosophical Quarterly* **82**: 1–10.

Chaitin, G. 2006. *Meta Math: The Quest for Omega*. New York: Vintage Books.

Collins, H. 1985. *Changing Order: Replication and Induction in Scientific Practice*. Chicago, IL: University of Chicago Press.

Collins, R. 1994. "Against the Epistemic Value of Prediction Over Accommodation". *Noûs* **28**: 210–24.

Devitt, M. 1991. *Realism and Truth*. Oxford: Blackwell.

Drell, S. D. 1979. "Experimental Status of Quantum Electrodynamics". *Physica A: Theoretical and Statistical Physics* **96**: 3–16.

Einstein, A. 1920. *Relativity: The Special and General Theory*. London: Methuen.

Feyerabend, P. 1976. "On the Critique of Reason". In *Method and Appraisal in the Physical Sciences*, C. Howson (ed.), 309–39. Cambridge: Cambridge University Press.

Fine, A. 1986. "Unnatural Attitudes: Realist and Instrumentalist Attachments to Science". *Mind* **95**: 149–79.

Friedman, F. L. 1966. *Physics*. Boston, MA: D. C. Heath.

Frisch, D. H. & J. H. Smith 1963. "Measurement of the Relativistic Time Dilation Using Mesons". *American Journal of Physics* **31**: 342–55.

Glymour, C. 1980. *Theory and Evidence*. Princeton, NJ: Princeton University Press.

Hafele, J. S. & R. E. Keating 1972. "Around-the-World Atomic Clocks: Predicted Relativistic Time Gains and Observed Relativistic Time Gains". *Science* **177**: 166–70.

Hempel, C. 1966. *Philosophy of Natural Science*. Englewood Cliffs, NJ: Prentice Hall.

Jefferys, J. & W. Berger 1992. "Sharpening Ockham's Razor on a Bayesian Strop". *American Scientist* **89**: 64–72.

Junk, R. 1956. *Brighter Than a Thousand Suns*. London: Gollancz & Hart Davis.

Kitcher, P. 1992. "The Naturalists Return". *The Philosophical Review* **101**: 53–114.

Kronig, R. & V. F. Weisskopf 1964. *Collected Scientific Papers by Wolfgang Pauli*. New York: Wiley Interscience.

Lakatos, I. 1970. "Falsification and the Methodology of Scientific Research Programmes". In *Criticism and the Growth of Knowledge*, I. Lakatos & A. Musgrave (eds), 91–196. Cambridge: Cambridge University Press.

Lamb, D. 1991. *Discovery, Creativity and Problem-Solving*. Aldershot: Avebury.

Laudan, L. 1984. "Explaining the Success of Science: Beyond Epistemic Realism and Relativism". In *Science and Reality: Recent Work in the Philosophy of Science*, J. T. Cushing, E. McMullin & C. F. Delaney (eds), 83–105. Notre Dame, IN: University of Notre Dame Press.

Lewis, D. 1983. "New Work for a Theory of Universals". *Australasian Journal of Philosophy* **61**: 347–77.

Lyons, T. 2002. "Scientific Realism and the Pessimistic Meta-Modus Tollens". In *Recent Themes in the Philosophy of Science: Scientific Realism and Common-Sense*, T. Lyons & S. Clarke (eds), 63–90. Dordrecht: Kluwer.

Meixner, J. & G. Fuller 2008. "BonJour's *A Priori* Justification of Induction". In *Pre-Proceedings of the 26th International Wittgenstein Symposium*, S. Kostenbauer (ed.), 227–9. Kirchberg am Wechsel: Wittgenstein Society.

Mendel, G. 1913. "Experiments in Hybridisation". In *Mendel's Principles of Heredity*, W. Bateson (ed.), 7–48. Cambridge: Cambridge University Press.

Newton, I. 1960. *Sir Isaac Newton's Mathematical Principles of Natural Philosophy and His System of the World*, A. Motte (trans.). Berkeley, CA: University of California Press.

Psillos, S. 1999. *Scientific Realism: How Science Tracks the Truth*. London: Routledge.

Putnam, H. 1979. *Mathematics, Matter and Method*. Cambridge: Cambridge University Press.

Putnam, H. 1981. *Reason, Truth and History*. Cambridge: Cambridge University Press.

Reines, F. & C. L. Cowan 1953. "Detection of the Free Neutrino". *Physical Review* **92**: 830–31.

Salmon, W. 1971. *Statistical Explanation and Statistical Relevance*. Pittsburgh, PA: University of Pittsburgh Press.

Salmon, W. 1975. *Space, Time and Motion*. Encino, CA: Dickenson.

Stanford, P. K. 2000. "An Anti-Realist Explanation of the Success of Science". *Philosophy of Science* **67**: 266–84.

Stove, D. 1986. *The Rationality of Induction*. Oxford: Clarendon Press.

Van Fraassen, B. 1980. *The Scientific Image*. Oxford: Clarendon Press.

Watkins, P. 1986. *The Story of the W and Z*. Cambridge: Cambridge University Press.

Will, C. M. 1946. *Was Einstein Right?* New York: Basic Books.

Will, C. M. 1987. "Experimental Gravitation from Newton to Einstein". In *Three Hundred Years of Gravitation*, S. W. Hawking & W. Israel (eds), 80–127. Cambridge: Cambridge University Press.

Worrall, J. 1989. "Structural Realism: The Best of Both Worlds?". *Dialectica* **43**: 99–124.

Worall, J. 1999. "Two Cheers for Naturalised Philosophy of Science – Or: Why Naturalised Philosophy of Science is Not the Cat's Whiskers". *Science and Education* **8**: 339–61.

Wright, J. 1991. *Science and the Theory of Rationality*. Aldershot: Avebury.

Index